Managing Health Promotion Programs

Bradley R.A. Wilson, PhD
University of Cincinnati
Cincinnati, Ohio

Timothy E. Glaros, MA
Creative Business Consulting
Grey Eagle, Minnesota

Human Kinetics

Library of Congress Cataloging-in-Publication Data

Wilson, Bradley R. A., 1954-
Managing health promotion programs / Bradley R.A. Wilson, Timothy E. Glaros.
p. cm.
Includes index.
ISBN 0-87322-611-9
1. Health promotion. 2. Industrial hygiene. I. Glaros, Timothy E., 1945- . II. Title.
RC969.H43W54 1994
613'.068--dc20 93-38014
CIP

ISBN: 0-87322-611-9

Acquisitions Editor: Rick Frey, PhD; **Developmental Editor:** Mary E. Fowler; **Assistant Editor:** Anna Curry; **Copy Editor:** Elisabeth Boone; **Proofreader:** Kathy Bennett; **Indexer:** Theresa J. Schaefer; **Typesetter and Text Layout:** Julie Overholt; **Text Designer:** Keith Blomberg; **Interior Art:** Tom Janowski and Kathy Boudreau-Fuoss; **Cover Designer:** Jack Davis; **Printer:** Braun-Brumfield

Printed in the United States of America 10 9 8 7 6 5 4 3 2 1

Human Kinetics
P.O. Box 5076, Champaign, IL 61825-5076
1-800-747-4457

Canada: Human Kinetics, Box 24040,
Windsor, ON N8Y 4Y9
1-800-465-7301 (in Canada only)

Europe: Human Kinetics,
P.O. Box IW14, Leeds LS16 6TR, England
0532-781708

Australia: Human Kinetics, P.O. Box 80,
Kingswood 5062, South Australia
618-374-0433

New Zealand: Human Kinetics,
P.O. Box 105-231, Auckland 1
(09) 309-2259

This text is dedicated to

the memory of
Gene A. Wilson
and
James and Lillian Glaros

and

the healthy futures of
Ben, Steph, Alexa, and Nicholas

Contents

Preface

Sir David Steel, chairman of the British Petroleum Company from 1975 to 1981, once said, "Managers may be born, but they also have to be made." Traditionally, health promotion professionals have come from a background of health education, physical education, exercise physiology, or community health. The profession attracts people who have adopted healthy lifestyles and want to share their philosophies and practices with others. They often possess more technical knowledge than their program participants will ever need.

In an effort to share their knowledge with others, health professionals have been creating forums in the workplace. In the course of their career growth, many are promoted into management and supervisory positions and find their business knowledge insufficient to sustain them in their new roles. Employees trained in a health profession are suddenly business people. Unless they have been fortunate enough to have obtained some business training along the way, they may feel like someone thrown into deep water without knowing how to swim.

Managing Health Promotion Programs attempts to offer some swimming lessons. It is not intended to substitute for tertiary business programs. The compilation of practical information derived from our experiences in moving from "missionary" to "mercenary" simply provides a framework on which students and health, fitness, and wellness professionals new to management can build their skills. It offers the essential elements managers need to function in that role. We hope it will help make managers.

We begin by presenting an overview of health promotion, making a case for the role of business knowledge and skills in our profession. Part I begins with business policy and ethics, and Part II focuses on management. Portions are theoretical in nature, which is necessary to building a foundation for the many practical applications that are offered, including how to use an organizational chart, develop a job description, and integrate health promotion operations into an organization.

From management we move to marketing in Part III. After introducing traditional marketing principles, we present a practical approach to the marketing process. To meet the needs of worksite health promotion programs and vendors of health promotion services, we consider marketing programs internal and external to the organization. Part IV addresses the financial operations of the organization, discussing the internal and external economic forces that influence the organization as well as budgeting and purchasing. In Part V we examine the complete planning process. We consider long-range plans, project plans, work plans, and progress reporting.

We have chosen to use the term *organization* because health promotion professionals work in many different settings, including corporations, hospitals, government agencies, educational institutions, and not-for-profit agencies. Many of the examples use the worksite as the point of program delivery. Whatever the setting, the business concepts presented apply.

We hope that you will find *Managing Health Promotion Programs* to be an outstanding guide to managing effective health promotion programs. Our goal is to be practical, informative, and stimulating as we help you to understand and apply the business principles that will help make you a successful health promotion manager.

Acknowledgments

As is always true for a project like this, we could not have done all the work ourselves. We very much appreciate the input we have received and permission to use photos and documents as realistic and accurate examples. Special thanks in this regard are extended to our colleagues at Apple Computer, Ceridian (formerly Control Data), Dow Chemical, Fitness Systems, John Harris and Associates, and Tenneco. We also thank Linda Glaros for sharing her expertise in the written word by editing this text when it was in its rudimentary stage.

Photo Credits

Pages 57, 107, and 169 courtesy of The Dow Chemical Company.

Pages 1, 17, 23, 49, 73, 79, 101, 113, 133, 143, 153, 189, and 199 courtesy of Apple Computer, Inc., Peter N. Fox, photographer.

Pages 89, 173, and 219 courtesy of Apple Computer, Inc., John Greenleigh, photographer.

Page 225 courtesy of Apple Computer, Inc., Julie Chase and John Greenleigh, photographers.

Pages 13, 37, 123, and 207 courtesy of Tim Glaros.

Credits

Figure 1.1—From *Wellness Workbook* (p. xvi) by Travis and Ryan, 1988, Berkley, CA: Ten Speed Press. Copyright 1988 by John W. Travis, M.D. Reprinted by permission.

Figure 1.2—Reproduced with permission. *Heart and Stroke Facts: 1994 Statistical Supplement*, 1993 Copyright American Heart Association.

Figure 1.3—Reproduced with permission. *Heart and Stroke Facts: 1994 Statistical Supplement*, 1993 Copyright American Heart Association.

Table 1.1—From T. Golaszewski. (1992). Unpublished raw data. Used with permission.

Figure 4.1—Adapted by permission of John Harris and Associates, Inc.

Figure 4.2—Adapted by permission of Fitness Systems, Inc.

Figure 6.2—Adapted by permission of Fitness Systems, Inc.

Chapter 1

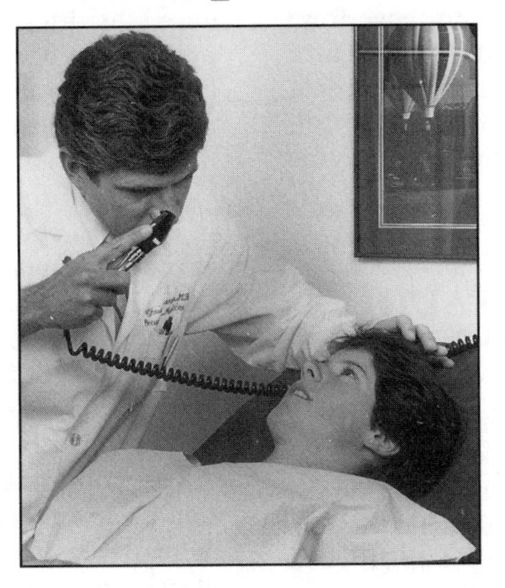

Introduction to Health Promotion

The field of health promotion is relatively new; it does not yet have a well-established tradition. Although at times this makes it difficult to clearly show all of the benefits health promotion has to offer, it also makes the profession very exciting. The field changes rapidly, and there is freedom to try new ideas. Skillful, energetic, and resourceful health promotion professionals can easily find opportunities to be creative and influential and to move into leadership positions. Precisely because health promotion professionals can soon become managers, interact with upper management, and be more autonomous than professionals in the more traditional business fields, they need the preparation of business training as well as health training.

HEALTH

Many definitions of health have been used in the last 50 years. The World Health Organization (WHO) developed a comprehensive definition in the 1940s, describing health as the "state of complete physical, emotional, and social well-being, not merely the absence of disease or infirmity" (Russell, 1975). More recently, the definition has been modified by others to accommodate the criticism that practically no one is completely healthy according to the WHO definition. The term *state* also implies that health is something that cannot be changed. These two problems resulted in a more comprehensive definition that is best shown in diagrammatic form (see

Figure 1.1). The wellness–illness continuum clearly shows that there are degrees of health and that it is possible to move from one level to another. The function of health promotion professionals is to keep people moving to the wellness side of the continuum, with the goal of seeing as many as possible at the far right.

ILLNESS AND HEALTH PROMOTION

At the extreme end of the illness side of the wellness–illness continuum is death. In the United States the leading cause of death is diseases of the heart and blood vessels (American Heart Association, 1992). Figure 1.2 shows that in 1993 nearly twice as many people died from cardiovascular diseases as from the second leading cause of death, cancer. Cardiovascular diseases are a major health problem in many other countries as well (see Figure 1.3).

Other leading causes of death in the United States are accidents, pulmonary diseases, pneumonia, influenza, suicide, and AIDS. All of these health problems contribute to the increasing costs of health care, and all are problems that can be remedied. For example, cardiovascular diseases are among those considered to be most influenced by lifestyle, and health promotion professionals can help people alter their lifestyles through such activities as exercise, weight loss, stress management, blood pressure control, and smoking cessation.

The need for health promotion services in the U.S. should continue to rise in light of the aging population and soaring health care costs. Because older individuals place heavy demands on the health care system and the cost of highly technical health care is spiraling upward, the use of health promotion must be given more consideration.

The potential benefits of health promotion programs go far beyond the reduction of health care costs to include decreased use of sick time and improved worker productivity, organizational image, and employee morale. Although these benefits are not easily measurable, many organizations, communities, and individuals are continuing to decide to use health promotion programs.

Health promotion services are offered in many settings, including worksite, community, commercial, educational, and clinical. All are important delivery points because they target different populations. Worksite programs reach employees and

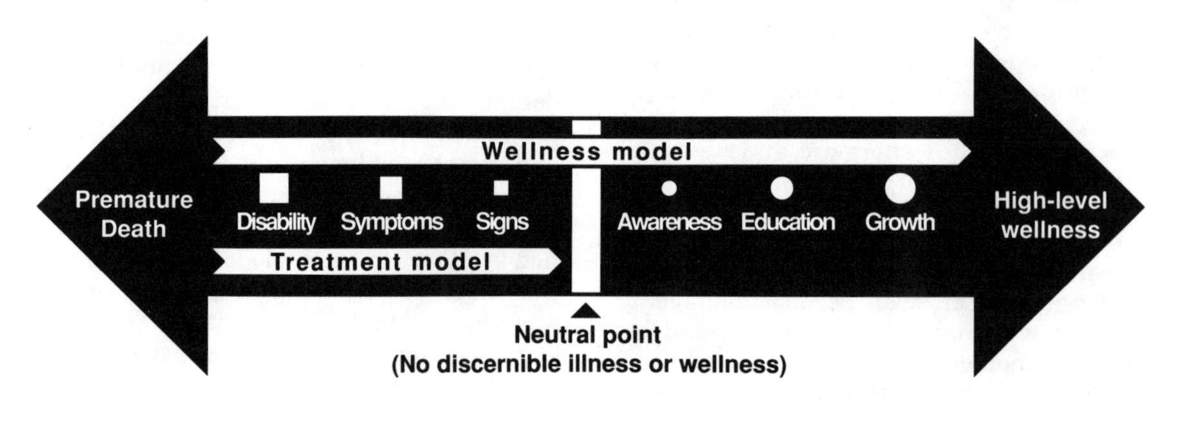

Figure 1.1 The health of a person is a point on the wellness-illness continuum.

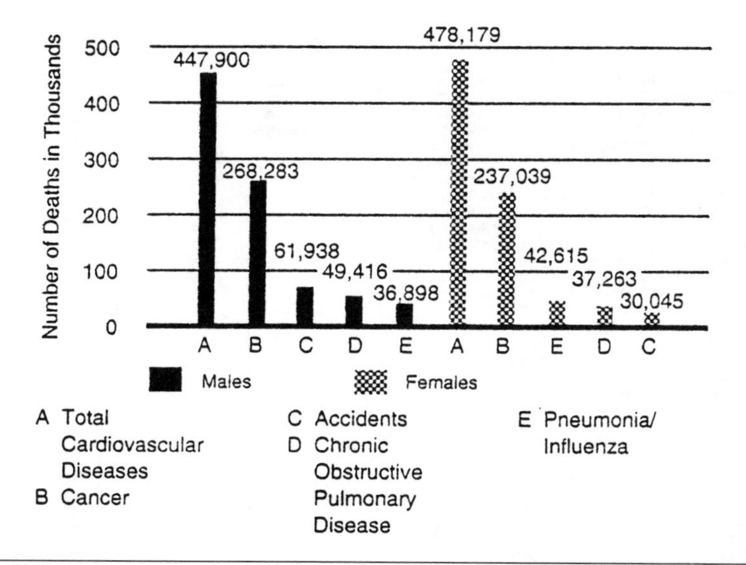

Figure 1.2 The leading causes of death for males and females in the United States. Reproduced with permission. *Heart and Stroke Facts: 1994 Statistical Supplement*, 1993 Copyright American Heart Association.

sometimes retirees and employees' families. Community programs reach the elderly, the unemployed, and others who do not have access to worksite programs. Commercial programs are available to individuals who are able to pay for them. Educational programs address students and young children. Clinical programs reach populations with specific needs. Together these delivery points can reach most people. However, in the future each individual delivery point must be expanded to meet everyone's needs.

FUTURE TRENDS

The health promotion field has changed dramatically in the last 10 years. More programs are being offered in more and different settings. Programs now address such topics as AIDS education, time management, and prenatal care. Health promotion professionals must monitor medical and social trends so they can provide appropriate programming to meet future needs.

Although it is impossible to predict the future accurately, we can consider history and trends in developing projections. Shifts in demographics can be predicted with some accuracy, although the implications of these changes are more difficult to predict. Nevertheless, demographic information is useful for long-range planning.

Trends in Demographics

Significant changes in U.S. demographics are expected by the year 2000. For example, the population is expected to rise 7% to 270 million people (Spencer, 1989). During this time the proportion of non-Hispanic whites is expected to drop from 76% to 72% (Spencer, 1989). The Hispanic-American population will increase to 31 million, or more than 11% of the population (Spencer, 1986). The African-American population will increase to 35 million, or more than 13% of the population (Spencer, 1989). Other ethnic groups,

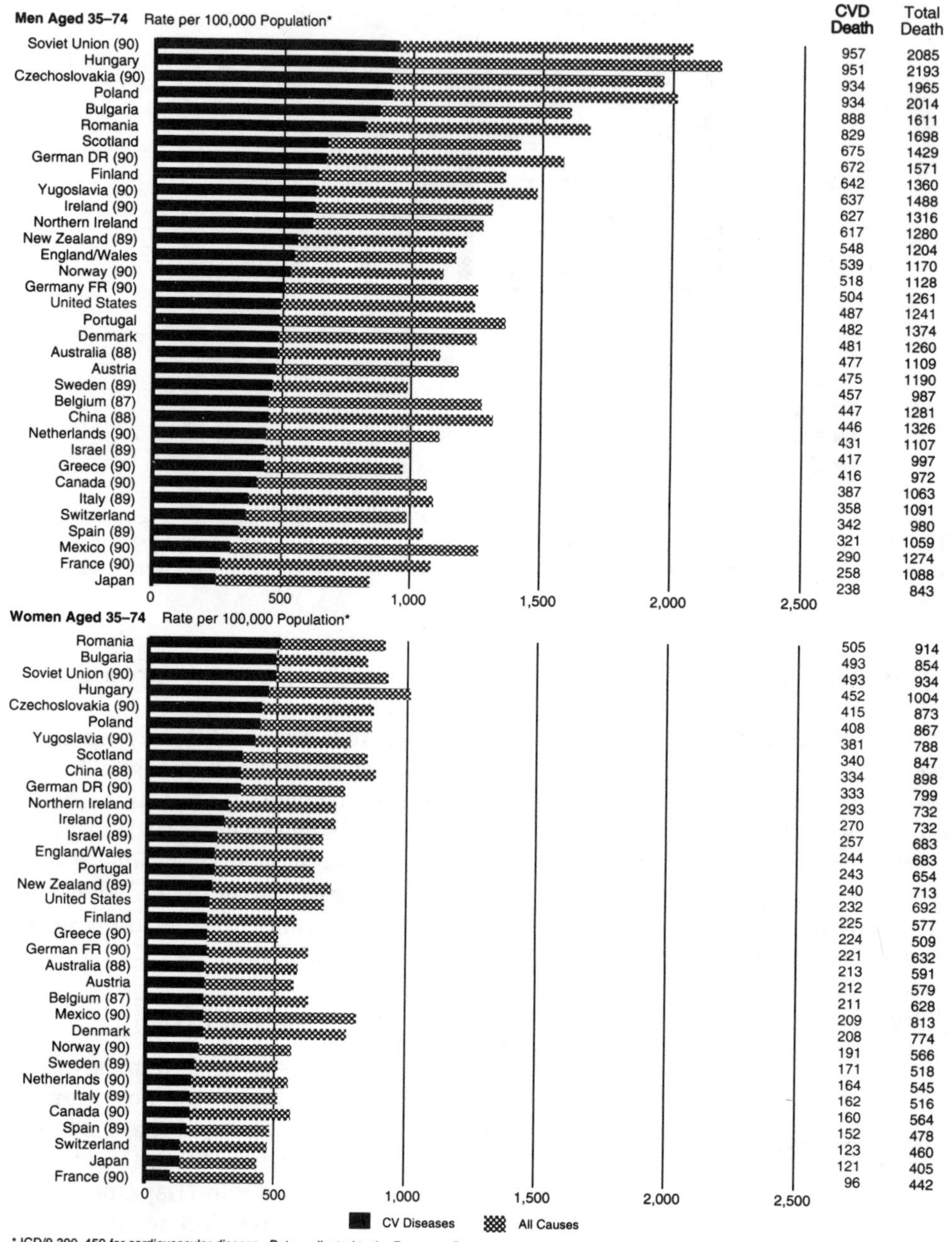

Figure 1.3 Death Rates for Cardiovascular Diseases and All Other Causes in Selected Countries, 1991 (or most recent year available).

Reproduced with permission. *Heart and Stroke Facts: 1994 Statistical Supplement*, 1993 Copyright American Heart Association.

including Native Americans and Asian-Americans, will continue to grow (Spencer, 1989), and health programs must be developed that address the specific needs of these expanding segments of the population.

Important changes also are expected with respect to age. By the year 2000 the median age in the United States will pass 36 years (Spencer, 1989). The number of children under the age of 5 will drop while the number of people over age 65 will increase to 35 million, representing 13% of the total population. The number of people over age 85 will also increase, reaching 4.6 million. The aging population will present new problems that the health promotion field will need to address.

Trends in Health Care

By the year 2000, the biggest problem related to health care in the United States will be cost. Although health care technology has improved dramatically over the last 30 years, the cost of care has increased to the point where many people cannot afford it.

Fortunately, many of the health problems that contribute to rising costs are largely preventable, such as heart disease, stroke, some types of cancer, many injuries, some cases of AIDS, alcohol and drug abuse, and inadequate immunizations. This has led the U.S. Department of Health and Human Services to conclude that "mobilizing the considerable energies and creativity of the Nation in the interest of disease prevention and health promotion is an economic imperative" (Public Health Service, 1990). The Healthy People 2,000 worksite goals appear on page 6.

Trends in Business

Some of the projected changes in U.S. business in the year 2000 will be related to demographic changes. The total work force will be larger (Bezold, Carlson, & Peck, 1986); however, there will likely be a worker shortage for entry-level jobs because of the low birth rates in the 1960s and 1970s (Taylor, 1991). Also, because of the changes noted previously in the population's ethnic makeup, the level of diversity in the work force will increase.

Technology will change the way business is conducted in the future (Taylor, 1991). Robotics, computer-aided design and manufacturing, and computer information systems will accelerate the pace of technological development. The ability to manage information and keep abreast of new technologies will be essential for success in the business world.

CHANGES IN HEALTH PROMOTION

Changes in demographics and business will result in changes in the health promotion field. A study by Miller and Tricker (1991) evaluated the expectations of 76 health and fitness professionals. Those surveyed anticipate many changes in health promotion. Respondents see an increased need to develop programs for the elderly, retirees, children, women, and hourly workers. Decreases are expected in health promotion services targeted to heart/lung disease patients, athletes, and individuals who are obese or have eating disorders. The importance of targeting employees of large organizations will change little (but will still be considered important) while the importance of targeting employees of small organizations will increase.

Miller and Tricker (1991) also evaluated predicted increases in staff sizes in different program settings, with the largest increase predicted for business and industry. Increases are also expected for medically oriented private clinics, community programs,

Healthy People 2000 Worksite Goals

Increase the proportion of worksites offering employer-sponsored physical activity and fitness programs as follows:

20% of worksites with 50-99 employees
35% of worksites with 100-249 employees
50% of worksites with 250-749 employees
80% of worksites with greater than 750 employees

Increase to at least 50% the proportion of worksites with 50 or more employees that offer nutrition education and/or weight management programs for employees.

Increase to at least 75% the proportion of worksites with a formal smoking policy that prohibits or severely restricts smoking at the workplace.

Extend the adoption of alcohol and drug policies for the work environment to at least 60% of the worksites with 50 or more employees.

Increase to at least 40% the proportion of worksites employing 50 or more people that provide programs to reduce employee stress.

Increase to at least 85% the proportion of the workplaces with 50 or more employees that offer health promotion activities to their employees, preferably as part of a comprehensive employee health promotion program.

Increase to at least 20% the proportion of hourly workers who participate regularly in employer-sponsored health promotion activities.

Increase to at least 70% the proportion of worksites with 50 or more employees that have implemented programs on worker health and safety.

Increase to at least 50% the proportion of worksites with 50 or more employees that offer back injury prevention and rehabilitation programs.

Increase to at least 50% the proportion of worksites with 50 or more employees that offer high blood pressure and/or cholesterol education and control activities to their employees.

(From *Healthy People 2000*, 1990)

health or fitness clubs, insurance companies, universities and other education facilities, and hospitals.

Although staff size is predicted to increase in most settings, increases will not take place without reason. Therefore, managers must continually evaluate the effectiveness and benefits of health promotion to justify their programs to upper management and clients.

FUTURE PREPARATION AND CONTINUING EDUCATION

As the field of health promotion changes, both established professionals and students must acquire current knowledge and skills to be competitive in a rapidly changing profession.

Health Content

The health field changes continually: New needs arise, and some old needs decline in priority. In the early 1980s AIDS was not a major concern. By the early 1990s it had become an important priority. Health promotion professionals must be prepared to address all the important health problems identified by the market and the government. This requires course preparation in existing health content areas and continued education in the content areas associated with new health problems. Comprehensive health promotion programming is vital to serving the population and to the growth and development of the health promotion field.

Business Administration

A recent evaluation of the job descriptions of 14 health promotion managers/directors found that many business-related tasks were included (Golaszewski, unpublished data, 1992). When common elements were collapsed into like constructs, six major groups were identified (see Table 1.1). Most positions included business-related tasks such as program management, communications, personnel management, program evaluation, and public relations.

In addition to a thorough understanding of health-related content areas, health promotion professionals must develop business skills and knowledge. Because staffs tend to be small, health promotion professionals have a good opportunity to become managers or directors in a relatively short period of time. This factor alone points out the value of business training.

Table 1.1
Job Functions Found in a Survey of 14 Health Promotion Manager Job Descriptions

Job function	Percent
Program management: planning, promoting, directing	100
Personnel management: staff supervision, staff training, vendor contracting	100
Evaluation: evaluating, recordkeeping	100
Written communication: employee communication, reporting, professional communication	79
Public relations: media communication, customer servicing, internal consulting	71
Service delivery: testing, counseling, instructing	50

There are several other reasons why business training is an important component of health promotion professionals' preparation.

- First is effective program operation. Many different types of health promotion programs can be conducted simultaneously. Directing them effectively requires good management skills. For instance, within one setting over a 1-week period, programs must be coordinated in nutrition, stress management, smoking cessation, and prenatal care, as well as fitness center operations. This includes planning, scheduling, budgeting, marketing, purchasing, and training.

• Second is interface with upper management. Whether the setting is community, worksite, hospital, or another, the health promotion area must function within a larger organization. The leaders in these organizations generally have good business skills. To compete with other components for resources, health promotion professionals also must display good business skills.

• Third is individual growth and development, or upward mobility within the organization. Advancement in most organizations eventually will lead to a management position. Lack of business knowledge and management skills can result in slower advancement.

For these reasons health promotion professionals must focus on business as well as health. It is the purpose of this text to provide health promotion professionals a basic understanding of and appreciation for applicable business concepts.

College students who plan to enter the field of health promotion should take business-related courses. Professionals who have graduated without adequate business knowledge should pursue it by attending conferences, seminars, workshops, and university courses in such areas as accounting, finance, marketing, and management.

Certification

Certification enhances education and demonstrates that someone has the knowledge and skills needed to enter or advance in the health promotion field. Many groups certify individuals in the broadly defined health promotion area. Some of these are

- the American College of Sports Medicine;
- the IDEA Foundation;
- the Aerobics and Fitness Association of America;
- the National Commission for Health Education Credentialing, Inc.; and
- the Aerobics Institute.

Most of these certifications are for specific components of health promotion. Students and professionals who are seeking certification should contact each group for detailed information. Before deciding to pursue a specific certification, follow these steps:

- Evaluate your own knowledge and skills.
- Determine your long-term and short-term goals.
- Assess the reputation of the certifying group.

These steps will guide you in making the best decision for your own needs.

SUMMARY

The field of health promotion has changed substantially in the last 10 years, and we can expect substantial changes in the future. Health promotion professionals must stay abreast of changes in the demographics of the work force, technology in the workplace, and the delivery of health care. In addition, professionals must pursue new knowledge and incorporate the information into effective programs, not only for their individual success but also for the success of the field.

Although health promotion skills will help secure a job, business skills are important for advancement. The remainder of the text addresses specific areas of business: business foundations and principles, management, marketing, financial operations, and planning. Developing knowledge and skills in these areas will substantially contribute to a successful career in health promotion.

SUGGESTED RESOURCES

American College of Sports Medicine. (1992). *ACSM's health/fitness facility standards and guidelines*. Champaign, IL: Human Kinetics.

American Heart Association. (1992). *1992 heart and stroke facts*. Dallas: Author.

Association for Fitness in Business. (1992). *Guidelines for employee health and fitness programs*. Champaign, IL: Human Kinetics.

Bezold, C., Carlson, R., & Peck, J. (1986). *The future of work and health*. Dover, MA: Auburn House.

Bialkowski, C. (1991). The future of corporate fitness. *Club Industry*, **7**, 33-38, 94-95.

Donatelle, R., Davis, L., & Hoover, C. (1991). *Access to health*. Englewood Cliffs, NJ: Prentice-Hall.

Downie, R., Fyfe, C., & Tannahill, A. (1990). *Health promotion: Models and values*. New York: Oxford University Press.

Institute of Medicine, Committee for the Study of the Future of Public Health. (1988). *The future of public health*. Washington, DC: National Academy Press.

Miller, C., & Tricker, R. (1991). Past and future priorities in health promotion in the United States: A survey of experts. *American Journal of Health Promotion*, **5**, 360-367.

Naisbitt, J., & Aburdene, P. (1990). *Megatrends 2000: Ten new directions for the 1990's*. New York: Morrow.

Public Health Service. (1990). *Healthy People 2000: National health promotion and disease prevention objectives* (PHS Publication No. 91-50212). Washington, DC: U.S. Government Printing Office.

Romer, K. (1987). Corporate health promotion in a post-industrial society: Circa 2000 A.D. *Health Education*, **18**, 16-20.

Russell, R. (1975). *Health Education*. Washington, DC: National Education Association.

Schwefel, D. (Ed.) (1987). *Indicators and trends in health and health care*. New York: Springer-Verlag.

Shephard, R. (1986). *Economic benefits of enhanced fitness*. Champaign, IL: Human Kinetics.

Sol, N., & Wilson, P. (Eds.) (1989). *Hospital health promotion*. Champaign, IL: Human Kinetics.

Stoto, M., Behrens, R., & Rosemont, C. (1990). *Healthy People 2000: Citizens chart the course*. Washington, DC: National Academy Press.

Taylor, R. (1991). Trends to watch in the 1990's. *Leadership*, 15-18.

World Health Organization. (1988). *Health promotion for working populations: A report of a WHO expert committee on health promotion in the work setting*. Geneva: Author.

Part I

Foundations and Principles

Business operations can be divided into several major disciplines: management, marketing, budgeting, and planning. However, there are considerations that affect all disciplines within business. These are addressed in Part I.

Part I begins with chapter 2, in which we discuss policies and procedures. Written policies and procedures guide decision making and are the foundation of business operations. An important related factor is business ethics. Chapter 3 presents an overview of ethical considerations in organizations. Statements of ethics also provide a guide for making decisions.

On completing Part I you should have a basic understanding of the foundations of business decision making. These principles will be important for all the decisions made in each discipline addressed in this text. With these foundations and principles at hand, Part II will discuss the major business areas, beginning with management.

Chapter 2

Policies and Procedures

Health promotion professionals are subject to policies and procedures, and they may be responsible for developing policies and procedures that are related to their specific operations and facilities. For example, a fitness facility must have policies related to

- who can use the facility,
- how certain equipment must be used to assure safety, and
- screenings needed before participation.

Clear policies on these and other issues are needed for health promotion programs to function at optimal levels. In many cases, specific procedures will further define the implementation of a particular policy.

PURPOSES OF POLICIES

For a policy to be useful, it should have two characteristics. First, it should relate to a problem or situation that arises more than once. Because policies are time-consuming to prepare, it is not cost-efficient to write them for unusual situations. Second, a policy should be a guide for decision making. Its purpose is not to specify a particular action, but to provide some guidelines for making decisions that will best serve the organization. Therefore, it is important that policies be written with the organization's objectives in mind.

POLICY DEVELOPMENT

In general, policy formulation is the responsibility of all managers in an organization.

Specific policies should be developed by the manager who is responsible for implementing the policy. Therefore, health promotion managers should be prepared to develop policies that are related to the health promotion program.

One method of policy development is the participative approach, which involves seeking input from employees who represent related departments and diverse groups in the organization. In worksite health promotion, this approach can be maximized by consulting with the employee advisory group. These representatives can assist managers in establishing policies that serve the interests of the participants as well as those of the program managers.

Within a health promotion program, policies can be departmental or organizational (see Table 2.1). Organizational policies are developed in conjunction with higher-level managers to determine how the health promotion functions fit in with the organization as a whole. These policies are likely to be included in the policy manual. Departmental policies are more specific to health promotion functions and are the responsibility of the health promotion staff. These policies will not be included in the organization's policy manual, and managers should cover them with each health promotion program participant during orientation.

Table 2.1
Departmental and Organizational Policies

Departmental policies

Registration/cancellation
Minimum education requirements for professional positions
Access to fitness facilities
Time limits for incentive redemption
Course/activity completion for incentive credit

Organizational policies

Eligibility for program
Solicitation on premises (e.g., by vendors at a health fair)
Conference room priority
Confidentiality of records (e.g., HRAs, blood tests)
Equal Employment Opportunity/affirmative action hiring

INTEGRATION OF POLICIES

If health promotion operations are a smaller component of a larger organization, as with worksite health promotion, the health promotion program must operate within the policies of that organization. In many cases these policies are administered by the human resources department. Specifically, they might include

- hiring, termination, and transfer practices;
- distribution of communications, such as newsletters and check stuffers;
- purchasing of goods and services; and
- budget guidelines.

In each of these situations managers should consult the appropriate policies before proceeding.

PUBLIC POLICY

In many situations policies must be developed to regulate organizations in the public interest. These policies, which are external to the organization, are called public policies. For example, public policies are typically needed to regulate

- what utility companies charge for their services,
- monopolies and other forms of unfair competition,

- trade, and
- product safety.

Public policies have several functions that are related to maximizing consumer benefit. First, they promote efficient management within the organization and promote the efficient use of limited resources. Public policies should result in less waste as a result of poor management and allow better allocation of limited resources within society. Second, public policies provide incentives to update technology to increase output. Third, they promote the equitable distribution of economic benefits to individuals. Fourth, they promote social values, resulting in decisions that may not be based on economic considerations but that provide individual employees better work environments and greater job satisfaction. For example, in the health promotion field, a public policy on weight-loss programs is needed to prevent unqualified individuals from offering programs and to restrict programs that are not based on proven scientific principles.

PROCEDURES

Most policies need guidelines that allow staff members to implement them. These guidelines are called procedures. For example, a policy may state that only enrolled, eligible employees may use the fitness facility.

To enforce this policy uniformly, the following procedures may be used:

- All enrolled employees must present their membership card to staff before being issued towels and lockers.
- Each membership card must be matched to eligibility and liability waiver records by being run through the computerized card reader.
- Employees without valid membership cards cannot enter the facility without approval of the fitness facility manager.
- A temporary membership card valid for 24 hours may be issued by the manager after another liability waiver has been signed and filed.

Procedures may be simple or complex, depending on the complexity of the related policy. Many simple policies do not require formal procedures because the implementation is obvious. More complex procedures may be revised from time to time to clarify aspects of the policy that may not have been anticipated initially. Managers should encourage staff members to suggest ways to simplify or improve departmental procedures.

SUMMARY

Policies provide a framework within which health promotion professionals operate. They range from broad public policies on health care and advertising to organizational policies followed by all departments and departmental policies that are developed and implemented by the health promotion staff. Managers may also need to develop specific procedures to clarify the implementation of some policies.

The subsequent parts of this text address the areas of management, marketing, budgeting, and planning. The policies and procedures discussed in this chapter, as well as the ethics considerations examined in the next chapter, influence each of these areas. We recommend that you keep the policies and procedures discussion in mind as you study the specific business operations addressed in the text.

KEY TERMS

Definitions of the following terms appear on the page shown in parentheses:

departmental policy (p. 14)

procedures (p. 15)

public policy (p. 14)

organizational policy (p. 14)

SUGGESTED RESOURCES

Butler, H. (1987). *Legal environment of business: Government regulation and public policy*. Cincinnati: South-Western.

Frederick, W., Davis, K., & Post, J. (1988). *Business and society: Corporate strategy, public policy and ethics*. New York: McGraw-Hill.

Luffman, G. (1987). *Business policy: An analytical introduction*. New York: Basil Blackwell.

Marcus, A., Kaufman, M., & Beam, D. (1987). *Business strategy and public policy*. New York: Quorum Books.

Page, S. (1984). *Business policies and procedures handbook: How to create a professional policies and procedures publication*. Englewood Cliffs, NJ: Prentice-Hall.

Shepherd, W. (1985). *Public policies toward business*. Homewood, IL: Irwin.

Chapter 3

Business Ethics

The ethics of an organization are part of its culture. If managers care about the employees, community, customers, and suppliers, the actions and attitudes of everyone in the organization will reflect this commitment to sound business ethics.

VALUES AND ETHICS

Ethics are principles of conduct that guide the actions and decisions of an individual or a group. They are based on a set of values that are learned from life experiences. The values of individuals in an organization influence their actions as decision makers and thus determine the values of the organization.

THE ORGANIZATION'S INFLUENCE ON SOCIETY

Organizations are powerful forces in society. They provide goods, services, jobs, income, and wealth. Both for-profit and not-for-profit entities influence society's values through the nature of the work environment and advertising they create. In the case of for-profit corporations, some people perceive a conflict between the corporation's desire to maximize profits and the need to achieve social objectives. Not-for-profit health agencies use advertising and educational programs that also influence society's values.

Government Regulation of Business

Many health promotion professionals work for profit-making organizations such as commercial health clubs and companies that manage health promotion programs for employers. In a free-enterprise system, profit is the reward for hard work and good management. Unfortunately, from an ethical standpoint, seeking profit also can be associated with cutting corners and abusing the system.

Abuses in pursuit of profits have led to increased government regulation of business. Initially, some business leaders opposed this regulation because it constrained their ability to earn profits. However, many came to realize that many forms of regulation are beneficial because they protect the ethical business against less ethical competitors. Regulatory examples include

- minimum-wage laws,
- occupational safety and health regulations,
- environmental legislation, and
- truth-in-advertising guidelines.

Today many businesses are looking beyond profits and toward increased social responsibility. For example, they are

- offering health promotion programs for their employees;
- providing better customer service;
- supporting community programs, such as the arts and outreach to the underserved; and
- contributing to nonprofit agencies and education.

As governments cut back on services to balance budgets and keep taxes down, for-profit organizations are stepping in and filling gaps in their communities.

BUSINESS ETHICS AND HEALTH PROMOTION

Health promotion professionals are often placed in situations where ethical dilemmas arise. Professionals may be exposed to insider information or confidential personal data about highly placed individuals, and there may be temptations to enter into subtle and direct conflicts of interest. Advance knowledge of the implications of such situations will prepare individuals to avoid personally and professionally damaging consequences within their organization and profession. Like policies and procedures, ethics should always be considered in decisions relating to health and to business.

The setting in which a health promotion professional works often determines the ethical issues that arise. Considerations may differ depending on the market an organization serves or on whether it is a for-profit or not-for-profit entity. Although numerous ethical situations can arise in any organization, some of the more common issues are confidentiality, vendor and purchaser relationships, advertising and promotion, conflicts of interest, proprietary information, and the quality of goods and services.

Confidentiality

Health promotion professionals frequently have access to personal information that should not be shared with others. For example, they may be aware of program participants who

- are taking medications, have physical problems, diseases, or other chronic conditions; or
- are pregnant.

Health promotion personnel need such information to give exercise prescriptions and other forms of counseling accurately. Breaches of confidentiality can cause the health promotion function to lose credibility. As a general rule, participant information should be shared only with appropriate health professionals unless laws mandate disclosure to others. If questions arise about responsibility concerning disclosure, the health promotion professional should seek legal advice.

Vendor and Purchaser Relationships

The relationship between vendors and purchasers has the potential for ethical abuse. Vendors have much to gain by making a sale, and many vendors offer incentives to individuals who make purchasing decisions. Incentives might include product samples, meals, tickets to sporting or cultural events, material gifts, and trips.

The ethical issues surrounding giving and accepting these incentives are complex. Organizations should have policies that direct their employees in the decision-making process. Health promotion professionals who deal with vendors must carefully consider appropriate policies and personal values. The decision to accept or offer incentives should be based on what is best for health promotion program participants and the organization representing them.

Advertising and Promotion

Attracting new members to a health promotion program or attracting new consumers of health promotion goods and services is an important aspect of the overall health promotion function. Advertising and promotions are typically used to accomplish these objectives. Ethical problems can arise if care is not taken to accurately represent the goods or services offered and to avoid offending any special groups within the population. The credibility of health promotion depends in large part on the ethical and value-driven use of advertising and promotion.

Conflicts of Interest

It is important for all professionals to be aware of situations where they have decision-making power or access to information that can be used to benefit them in other areas. This applies particularly to individuals who are working as independent consultants. Involvement in such programs as the American Heart Association's Heart at Work can give access to information about organizations that are considering the implementation of health promotion programs. If such information were used to target a particular organization for marketing services, it would constitute a conflict of interest.

Most organizations, including the American Heart Association, have policies to prevent conflicts from arising. Violation of such policies would create a serious ethical problem. Because volunteering is so important to our society, competing with volunteer agencies after obtaining information from them can damage the entire health promotion effort. When potential conflicts of interest arise, health promotion professionals should temporarily excuse themselves and decline to be involved in making decisions or sharing information.

Proprietary Information

Particularly in worksite health promotion, access to proprietary information can create problems. Because health promotion professionals have regular contact with a cross-section of employees, information not intended for the general employee population can be communicated in informal discussions. This information might concern sensitive negotiations or the decision to close a plant or downsize the work force. Health promotion professionals must recognize the sensitive nature of such information and keep it confidential.

Another ethical problem can arise when a health promotion professional learns that someone is violating policies. By reporting the violation one may lose credibility with the source, and by not reporting it one may lose credibility with management. This is a difficult situation that must be resolved on an individual basis.

Quality of Goods and Services

Perhaps the most important ethical concern of a health promotion organization is providing the best programs, goods, and services possible. This requires acting unselfishly to maximize the benefits to participants and consumers. This is the foundation of ethics for all health promotion professionals and is vital to the success of the health promotion field.

SUMMARY

High ethical standards are an essential element of effective health promotion efforts. Individuals and organizations can avoid breaches of ethical conduct by practicing

- confidentiality,
- ethical customer and vendor relationships,
- sound advertising and promotion,
- appropriate actions in conflict of interest situations,
- appropriate use of proprietary information, and
- continuous development of high-quality goods and services.

KEY TERMS

conflict of interest (p. 19)
ethics (p. 17)
values (p. 17)

SUGGESTED RESOURCES

Bick, P. (1988). *Business ethics and responsibility: An information sourcebook.* Phoenix: Oryx Press.

Cavanagh, G., & McGovern, A. (1988). *Ethical dilemmas in the modern corporation.* Englewood Cliffs, NJ: Prentice-Hall.

DesJardins, J., & McCall, J. (1990). *Contemporary issues in business ethics.* Belmont, CA: Wadsworth.

Donaldson, J. (1989). *Key issues in business ethics.* San Diego: Academic Press.

Doxiadis, S. (1987). *Ethical dilemmas in health promotion.* Chichester, NY: Wiley.

Freeman, R. (1991). *Business ethics: The state of the art.* New York: Oxford University Press.

Madsen, P., & Shafritz, J. (Eds.) (1990). *Essentials of business ethics.* New York: New American Library.

McHugh, F., & Reidenbach, R. (1988). *Business ethics: Where profits meet value systems.* Englewood Cliffs, NJ: Prentice-Hall.

Walton, C. (Ed.) (1990). *Enriching business ethics.* New York: Plenum Press.

Part II

Management

To this point we have been focusing on the foundations and principles for making business decisions. We have identified the major subjects for policies and procedures and business ethics. You now should be able to apply this information in the major disciplines of business.

In Part II we begin our journey through the business disciplines with a discussion of management. Because management, when broadly defined, can be part of all disciplines within business, we will address in Part II those functions concerned with the relationships among individuals and departments within the organization and the supervision of employees.

In chapter 4 we discuss the organization's structure, explaining how departments relate to one another and to upper management. In chapter 5 we consider the individual and how he or she relates to the organization. Chapters 6, 7, and 8 focus on the employment process, beginning with designing the job and continuing with the staffing process, appraising performance, and training and developing employees. In chapter 9 we discuss managing communications within the organization; in chapter 10 we consider the use of computers in management operations.

On completing Part II you should have a good management framework to apply to future chapters in the text. Part III continues our journey through the business disciplines with a focus on marketing of health promotion programs.

Chapter 4

The Organization

Whether health promotion is a small part of an organization or its main function, a solid understanding of organizations in general is essential for effective program management. In either situation the health promotion program should not be an independently functioning entity. Rather, it should be an important operating unit that complements all other activities in the organization.

ORGANIZATIONAL STRUCTURE

The structure of an organization determines the nature of operating units, the administrative hierarchy, reporting relationships, and formal lines of communication. Health promotion professionals must know how the organizational structure affects them and how to derive the maximum benefit from their position in the hierarchy. Where the health promotion functions are placed within the organizational structure determines the formal communication channels and ultimately the power and influence health promotion has in the organization.

Power and Authority

Power and authority differ in that power is defined as influence outside the organizational unit whereas authority is defined as influence within. In any case people with power or authority can have a major impact on the organization in terms of communication, job satisfaction, and the quality of interpersonal relationships.

More specifically, power increases the probability that the will of an individual will be carried out despite resistance. Formal authority is the right to provide positive or negative input to the behavior of others, such as promoting, demoting, hiring, firing, rewarding, and punishing.

Health promotion professionals should work continually to improve the skills necessary to obtain and maintain authority. Knowing the sources of power and authority gives professionals insight for making decisions that can enhance their effectiveness and the success of the health promotion effort.

Line and Staff

In any organizational structure, positions can be placed in one of two categories. *Line* positions refer to functions that are directly involved with the product or service ultimately produced by the organization. *Staff* positions refer to units that are service, supportive, or auxiliary in nature.

Depending on the type of organization, health promotion positions can be line or staff positions. In an organization where the health promotion department offers programs to its employees, the health promotion positions are clearly staff because they are providing technical advice and are not involved directly with the goods or services being produced. A different organization may be in the business of developing, implementing, and marketing health promotion services. There the positions in the health promotion department would be considered line positions because they are directly involved with services being produced and sold by the organization.

Whether health promotion positions in an organization are line or staff can affect the department's vulnerability to management's financial decisions. For instance, if a corporation is making cost-cutting decisions when sales are high, the cuts will more likely come from staff areas. If cost-cutting decisions are being made when sales are poor, the cuts will more likely come from line areas. Therefore, health promotion professionals need to make decisions that can stabilize the department during times of economic change.

Organizational Charts

The structure of an organization is portrayed in an organizational chart that depicts the formal pattern of relationships among the organization's various components (see Figure 4.1). Because these charts are oversimplified and concise, they can indicate only the major groupings of the organization's components. Also, organizational charts do not give complete information about formal reporting relationships and give essentially no information about informal relationships. Despite these shortcomings, organizational charts are useful management tools.

Organizational charts can be subdivided on the basis of several criteria. A chart that is subdivided into areas of activity depicts functional organization (see Figure 4.2). Other criteria used to subdivide an organizational chart are geographical area, products and services offered, customers served, technology, and work flow. Most large organizations use two or more of these subdivisions, and most use some form of functional organization (see Figure 4.3). Different subdivisions are found at different levels of the organizational chart.

Two other forms of organizational charts are project, as in Figure 4.4, and matrix, as in Figure 4.5. These are used primarily in organizations that are oriented toward several short, temporary projects as opposed to the development of a smaller number of

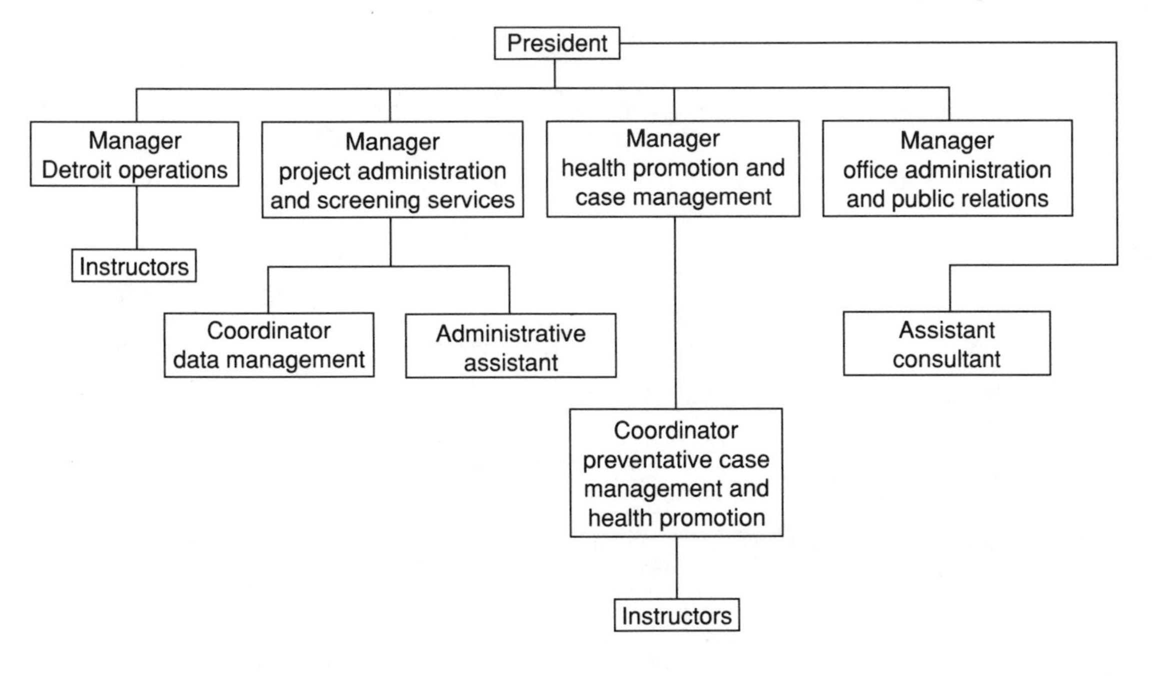

Figure 4.1 An organizational chart is used to identify the formal communication channels within an organization.

ongoing products and services. An organization with many small ongoing projects would use the matrix form, whereas an organization with a few major projects would use the project form. These forms of organization are appropriate for some health promotion consulting firms where many projects are being conducted for short periods of time, such as developing and implementing a 6-month smoking cessation program.

Project and matrix organizations are typically a combination of functional and product organizations and have three characteristics. They are

- frequently used with functional organization;
- composed of a team of individuals involved in the project who come from different functional areas; and
- composed of temporary organizational units.

These forms are appropriate for complex undertakings that require direct input from several different specialists.

Placement of Health Promotion Programs

The location of the health promotion department in the organizational chart is extremely important because it is the major indication of the formal authority associated with its leader and other personnel. Therefore, it is essential for health promotion managers to be familiar with the organizational chart and to understand the relationships among their area and other areas. Senior managers spend much time designing the organizational chart and considering input from interested parties.

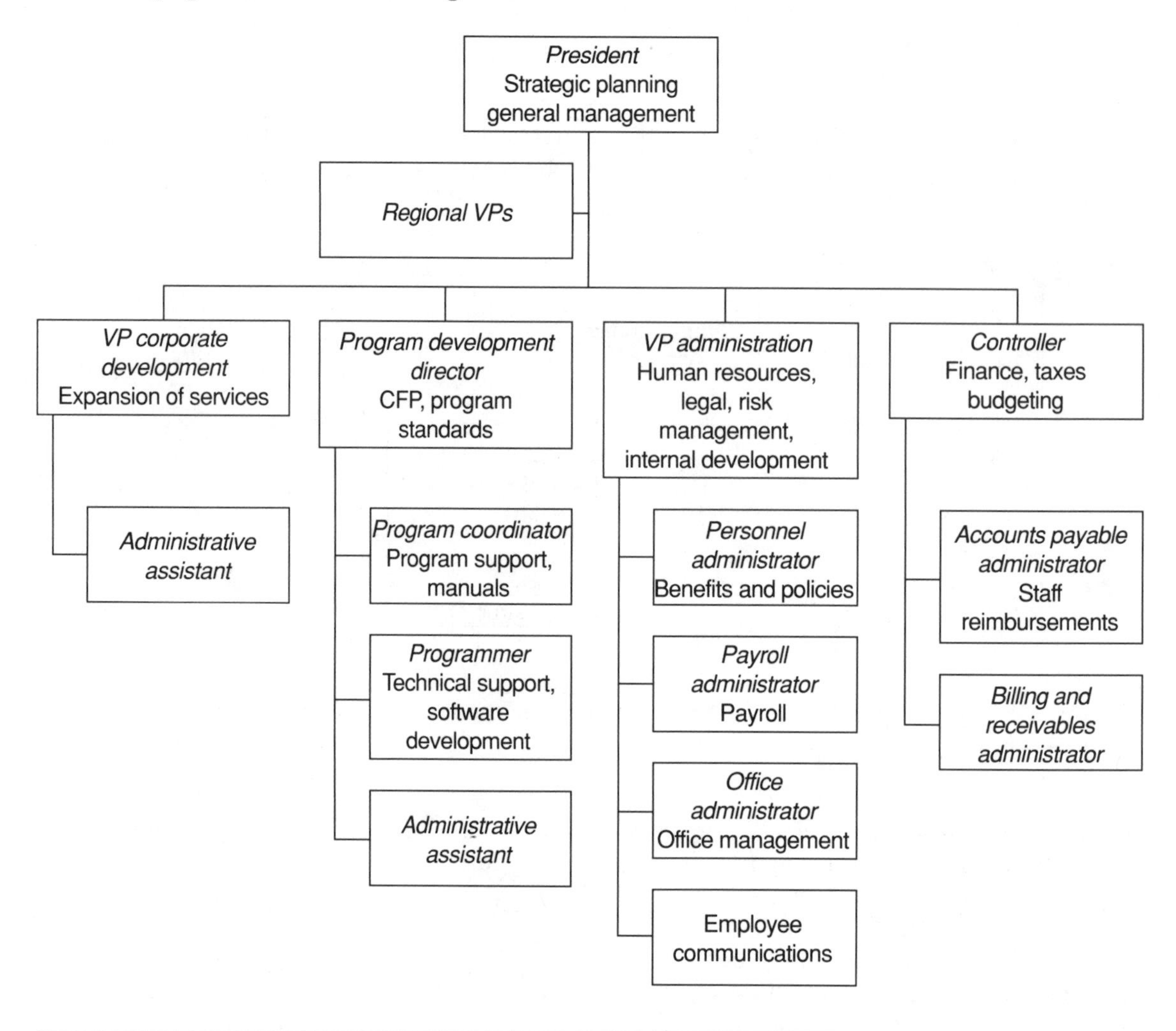

Figure 4.2 An organizational chart can be separated by job functions.

Health promotion professionals must be prepared to provide input to senior management that will enhance their power and authority and increase the impact of health promotion programs.

When a health promotion department offers programs to employees at the worksite, the department can be found in one of several locations on the organizational chart: human resources, personnel, employee assistance, employee benefits, or medical (see Figure 4.6). Although the ideal location depends on the organization, the best placement is directly under the chief executive officer.

However, reporting to the CEO is very unlikely. Reporting to a vice president is feasible and should be a goal for the health promotion department. This reporting relationship will give the department the influence to obtain necessary resources and to make favorable policy changes. In general it

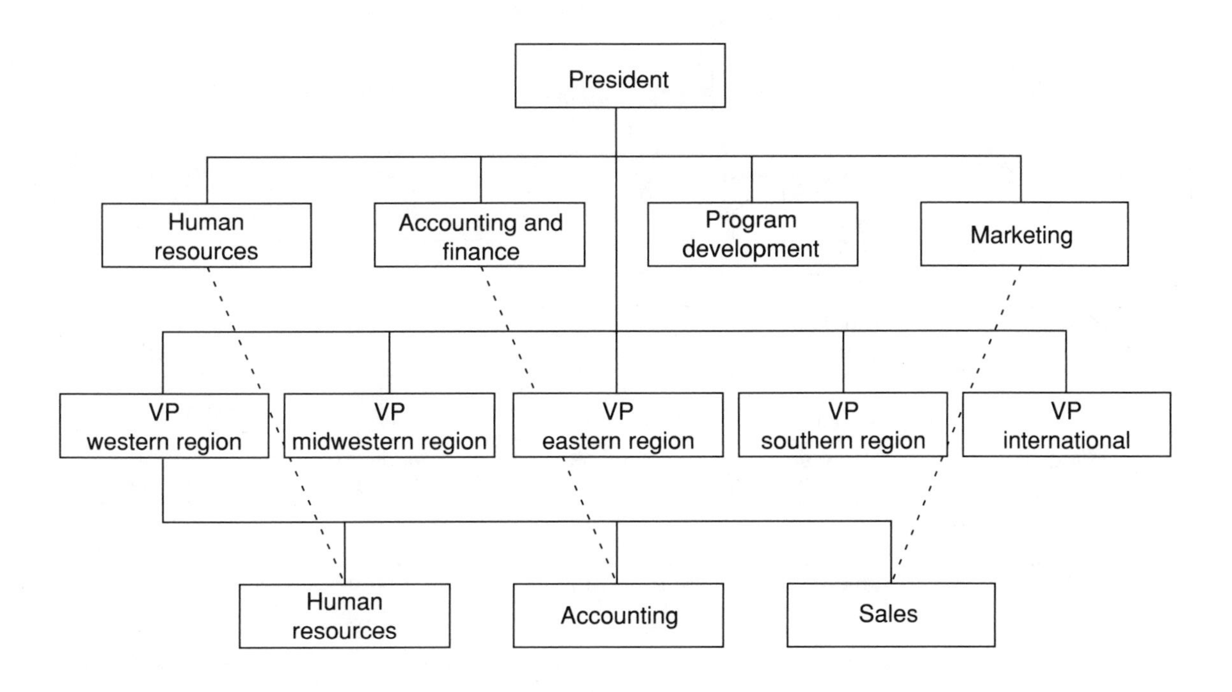

Figure 4.3 An organizational chart can be separated by geographical area.

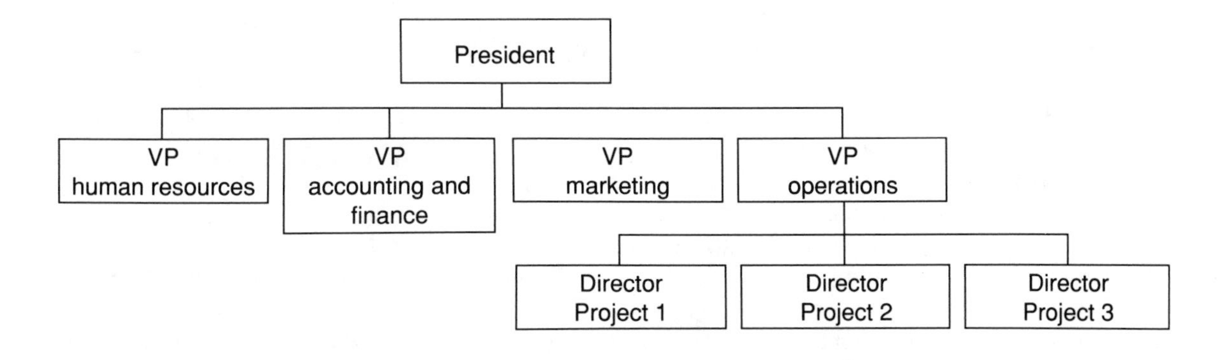

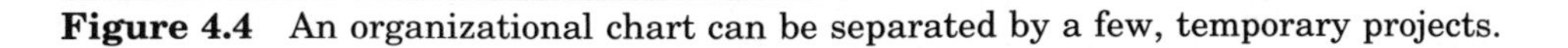

Figure 4.4 An organizational chart can be separated by a few, temporary projects.

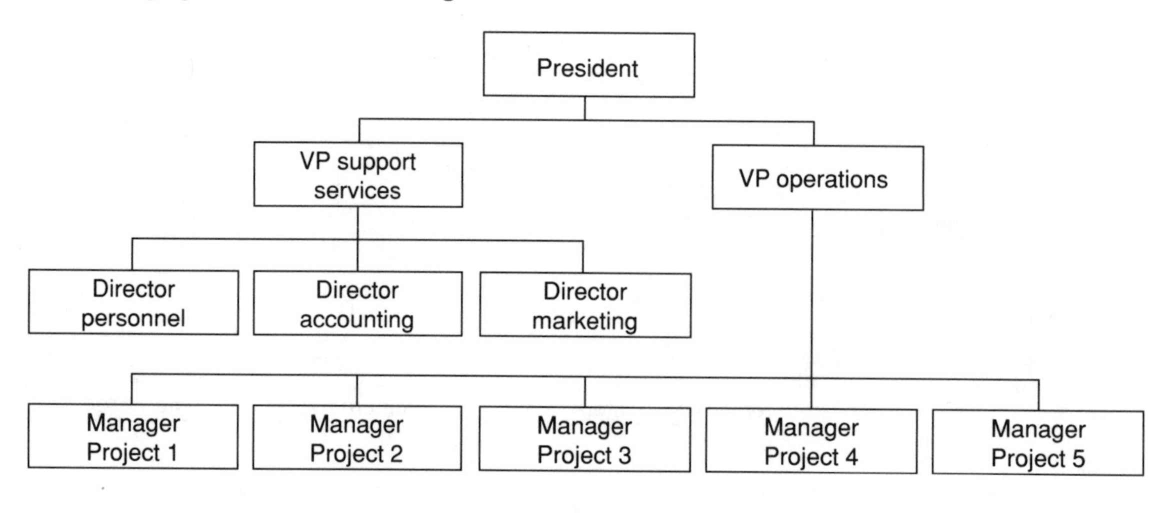

Figure 4.5 An organizational chart can be separated by many, temporary projects.

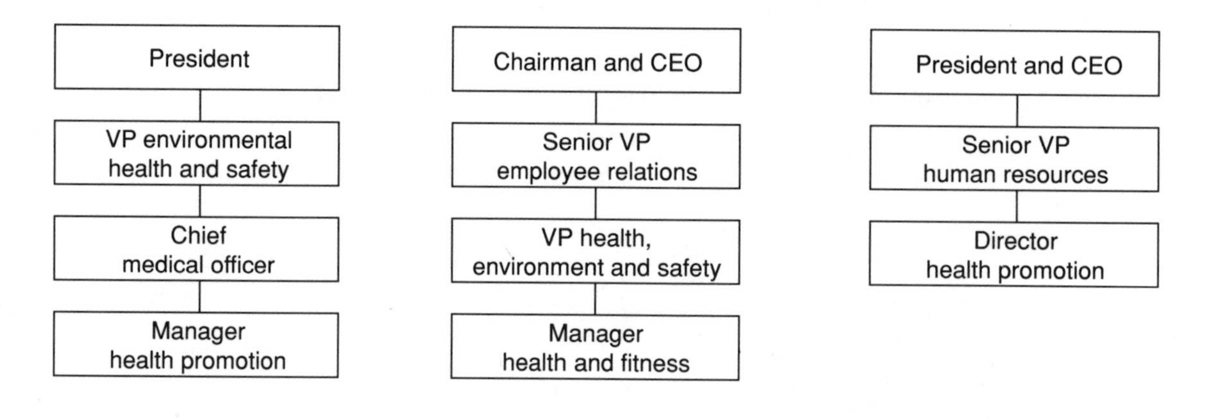

Figure 4.6 Locations of health promotion in the organizational chart.

is least advantageous to be located off to the side of a department (see Figure 4.7). This location provides minimal support from and formal power within the organization.

In organizations that provide health and fitness services to individuals outside the organization, the health promotion department is more likely to report directly to the chief executive officer. If it does not, it is more likely to be placed closer to the CEO because its functions are directly related to the services offered by the organization.

RELATIONSHIPS WITH UPPER MANAGEMENT

How an organization functions is a carefully planned process. Senior management spends considerable time making decisions that determine how the organization operates day to day and ultimately how successful it is. An understanding of these decisions will help health promotion professionals manage their own departments. If the department's goals

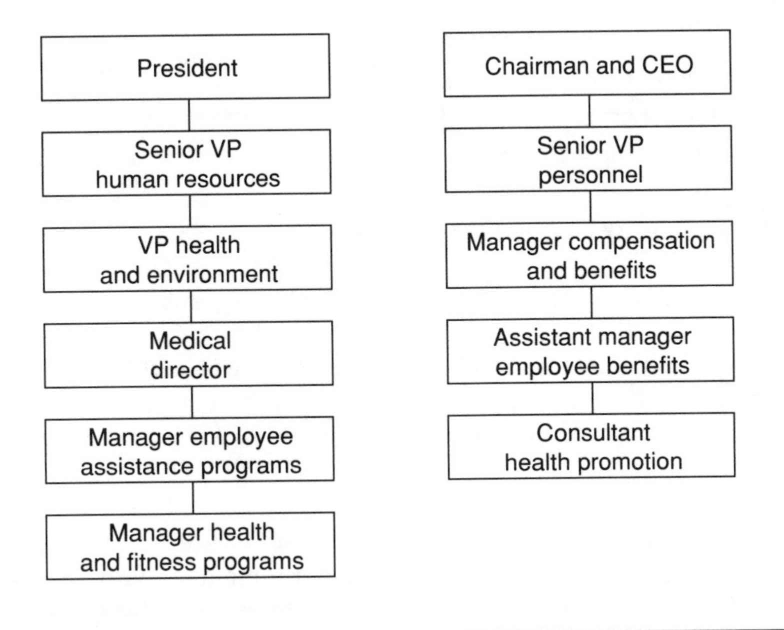

Figure 4.7 Poor location of health promotion in the organizational chart.

are consistent with those of the organization, the department will be in a better position to obtain resources and support from other components of the organization and to provide information for the development of organizational strategies.

Organizational Goals

Organizational goals are of two types (see Table 4.1). Official goals are formal written statements. These usually are very broad and provide little specific information about what strategies should be developed. Operational goals are goals employees attempt to achieve. However, operational goals are also nonspecific. If the organization is to attain its goals, all managers must understand the goals and agree on the specific strategies that must be implemented.

Organizational Performance and Goal Attainment

Senior management is ultimately responsible for the organization's performance and the attainment of its goals. An obvious goal for a corporation is to maximize profits. However, most organizations have additional goals, such as providing quality products and services, creating healthy work environments, and contributing to the community. To identify and establish goals, senior management must consider all interest groups. Some of the common interest groups for a health promotion service organization and their goals are shown in Table 4.2.

Organizational Strategy Development

Once an organization has set goals, managers must develop strategies to accomplish them. An organizational strategy is a detailed plan developed by senior management that is used as a basis for making decisions about how the organization should function and its relationship to outside parties. Ideally, the strategies are well defined, clearly documented, and used for

Table 4.1
Types of Organizational Goals

Official goals	Operational goals
To produce the highest quality products and services	To implement a total quality management program
To meet consumer demands	To establish ongoing customer user groups for feedback
To promote the health and well-being of all employees	To offer an employee health promotion program
To preserve the environment and conserve natural resources	To establish a recycling program throughout the organization

Table 4.2
Special-Interest Groups of Organizations

Interest group	Performance interest
Owners or shareholders	Profits
Managers	Attainment of goals
Suppliers	Timely payments
Consumers	Timely service
Creditors	Timely payments
Public	Job creation
Government	Affirmative action

making operating decisions. When these considerations are ignored, operating decisions are made individually and the direction from senior management is lost. Because this is likely to result in inefficiency and duplication of effort, all managers must understand the use of strategies.

Strategy management relies heavily on making accurate assessments of the organization's environment, including the economic situation, competition, suppliers, consumers, media, and government. Organizations obtain information about their environments in various ways. Large organizations have the resources and staff to support a specialized department that continually evaluates the environment and produces high-quality information. In contrast, small organizations are unable to allocate many resources to this area. These organizations frequently rely on information obtained from their day-to-day activities. Because most employees obtain information about their environment while performing their normal tasks, some information is readily available. However, this information must be processed and organized so it can be used for decision making.

RELATIONSHIPS WITH OTHER DEPARTMENTS

Within an organization are many departments that must work together informally to function effectively. The strength and quality of interdepartmental relationships depend on the kind of organization and how it is structured.

In a hospital or health promotion consulting firm where services are developed and marketed to other organizations, interdepartmental relationships generally are strong, because health promotion is a major operation of the organization. In an in-house health promotion department that provides services only to the employees of that organization, there may be fewer formal interactions with other departments. Most of the areas with which health promotion professionals are involved are related to human resources, and we will consider those areas in some detail in the following sections.

Compensation and Benefits

An important function of the human resources department is to establish appropriate compensation for employees at all levels and to design packages of fringe benefits. Three types of remuneration are:

- wages and salaries,
- fringe benefits, and
- nonmonetary rewards.

Examples of the different forms of remuneration are found in Table 4.3.

Table 4.3
Forms of Compensation

Monetary remuneration	Nonmonetary remuneration
Salary and wages	Special privileges
Time off with pay	Comfortable working environment
Insurance (life, health)	Feel appreciated by management
Retirement	Recognition for positive performance
Profit sharing	Job security
Tuition reimbursements	Feel challenged by job

Wages and Salaries

Administering wages and salaries includes many processes, most of which do not directly involve health promotion personnel. One important function of compensating personnel is to determine the relative value of each position to the organization. This process is used to establish the compensation level for each position and is based on job specifications. For this reason, health promotion managers must design job specifications that will allow the department to meet its objectives and demonstrate its value to the organization.

Fringe Benefits

Most organizations offer full-time employees a compensation package that includes supplementary or fringe benefits. Health promotion programs offered at the worksite are generally considered an employee benefit. Because their costs have increased substantially in recent years, fringe benefits have come under close scrutiny by management, with varying consequences for the health promotion field. On the positive side, health promotion programs have been implemented and expanded to help control the spiraling costs of health care, the most expensive fringe benefit. On the negative side, some programs have been considered expendable in organizations that are experiencing financial strain. The positioning of health promotion in a specific organization thus has a major impact on how this function should be managed with respect to the administration of benefits.

Management must examine the purposes of the fringe benefits program, including the health promotion program, and must evaluate the contribution of fringe benefits to the organization's financial objectives.

In general, there is a point where fringe benefits no longer contribute to the achievement of objectives and actually become a negative factor. At this point management may decide to allocate financial resources to rewarding superior performance instead of providing an across-the-board benefit. For this reason, managers must evaluate health promotion services in terms of their contribution to the attainment of financial objectives. This will give managers insight into the health promotion program's vulnerability to financial cutbacks and its value to the organization. Because no two situations are the same, each situation must be evaluated separately to reach informed judgments.

Nonmonetary Rewards

Financial rewards alone do not satisfy all employees' needs. A high salary with an excellent fringe benefit package is not adequate if the employee is under severe job stress, at high risk for a job-related accident, not accepted by co-workers, or not feeling fairly treated by management. Most employees need to feel personally fulfilled. Helping employees meet this need is the responsibility of all managers, including those in the health promotion department.

Medical Services

The medical departments of organizations vary widely in terms of the types of services offered and the number of staff members. Some organizations maintain an in-house staff of physicians, whereas others use outside physicians. Most large organizations, especially manufacturing organizations, have at least a registered nurse and a medical clerk.

Medical departments typically

- administer first aid;
- provide medical care;
- conduct medical exams (preemployment, periodic, and exit);
- dispense medications;
- provide technical advice on medical concerns;
- promote health and safety in the organization; and
- test employees for illegal drug use.

In some organizations the medical department is responsible for promoting employee health; in others the health promotion area is a separate unit. The formal relationships between the medical and health promotion functions should be defined in the organizational chart. However, many informal relationships can be developed for the mutual benefit of the two parties, and ultimately the organization.

For instance, the medical facilities can be excellent locations to inform employees of health promotion programs through literature and other materials. Also, medical personnel can provide referrals to health promotion programs. They are likely to know of people interested in losing weight, stopping smoking, or beginning exercise. They also know of people who are at risk, such as those who have high blood pressure. For these reasons, there should be strong informal communication between health promotion and medical professionals.

Employee Assistance

Employee assistance programs (EAPs) originally were created to deal with the problems of alcoholism. Today EAPs may also provide assistance with

- other forms of substance abuse,
- family problems,
- financial problems,
- legal problems,
- emotional problems, and
- eldercare and childcare issues.

Although professionals in EAPs perform some very different functions from professionals in health promotion, there are also many similarities. In general, EAP personnel are involved with immediate problems, such as emotional health, finances, drug and alcohol dependency, and family concerns. Their training is typically in counseling. Health promotion personnel are involved with long-term problems, such as coronary artery disease and cancer; their training is typically in health education and exercise. But because both kinds of professionals are involved with the health and welfare of employees, some joint programs may be warranted. For example, a stress management program offered jointly can benefit both departments and the organization.

Because the two groups have some common areas of concern, each can learn from the other. Of particular interest is management support. By learning why management supports an EAP, health promotion professionals can learn how to obtain more support for their programs. Also, each program should be able to help the other in areas like marketing and referrals. Organizations should have effective programs for both health promotion and employee assistance. Working together, the two disciplines can help achieve excellent employee health.

As organizations expand EAPs, health promotion and employee assistance come closer together. The result may be a combination of the two areas into one department. This has occurred in some organizations and can provide valuable opportunities for health promotion managers who are prepared to assume a leadership role over an expanded health program.

Labor Relations

An important consideration in many organizations is the management of employees who are unionized. Although unionized employees work in many job classifications, union roots are in labor-intensive jobs. For this reason, from a management standpoint, interaction with unions is commonly called labor relations.

If the health promotion program is intended to serve anyone who is a union member, managers must be aware of the state of union/management relations, the terms of the current contract, and the major contract items under discussion. Some important factors follow.

Health insurance benefits are a major item that is negotiated between unions and management. Health care is a basic need whose cost is increasing rapidly, and union leaders want to secure health care benefits for their membership. In organizations that offer health promotion programs, management may try to use health promotion as a bargaining tool against health insurance benefits. This tactic can be devastating to health promotion efforts because the union membership may develop a negative view of health promotion and refuse to participate in the programs. Ideally, health promotion services should be left off the bargaining table. It is the responsibility of health promotion managers to sell this idea to the organization's bargaining team.

Another major bargaining item is work time, including rest breaks, meal breaks, and overtime. Work schedules can affect employees' ability to take advantage of health promotion activities. Although health promotion managers have little influence on work time negotiations, being aware of them and planning ways to overcome problems can have positive results. For example, if a 30-minute lunch break is negotiated, most employees will not choose to exercise at that time. Therefore, managers may need to schedule exercise programs before and after work hours.

Major changes in the size of the work force also can affect the health promotion program. Layoffs, strikes, and lockouts are beyond the control of program managers.

However, when these events appear imminent, managers should develop a contingency plan to accommodate as many participants as possible and maximize the benefits from the resources that already have been committed to the program.

Occupational Safety and Health

In 1970 the federal government passed the Occupational Safety and Health Act (OSHA) to enforce safety and health standards in the workplace. This legislation placed an additional burden on employers to develop work rules, design jobs, and control the work environment to ensure the health and safety of employees. In most organizations this responsibility was given to safety personnel in the human resources area.

Although the functions associated with OSHA compliance and those associated with health promotion both concern the well-being of employees, the two disciplines are very different. In general, occupational safety and health focuses on the work environment, and many of the concerns have short-term implications. In contrast, health promotion focuses on lifestyle characteristics at and away from work, and most of the concerns have long-term implications. Nevertheless, in some organizations, especially smaller ones, these two areas may operate together. In organizations where they operate separately, joint programs can be offered. For instance, both departments have an interest in helping employees maintain healthy backs. This program could be taught by health promotion personnel and implemented by safety professionals.

From a cost justification standpoint, safety has a clear advantage over health promotion. An accident can happen in seconds and result in thousands of dollars lost. Lifestyle diseases can persist for years and also result in thousands of dollars lost. However, when program costs must be justified to management, the liabilities resulting from the lack of a safety program are much easier to identify than the liabilities associated with unhealthy lifestyles.

Training

Because health promotion efforts involve training individuals to live healthier lives, there is some overlap between the health promotion and training departments. A good way to integrate health promotion throughout the organization is to include some health education in every training session. Training sessions that last several days, such as sales meetings, can include exercise and relaxation breaks, healthy meals, and health screenings. Partial-day training sessions can include short relaxation breaks or nutrition breaks. One-hour sessions can begin with announcements of health program offerings or brief facts about health. Health promotion managers should be alert for opportunities to educate employees and make them aware of health promotion activities.

A responsibility related to training is new employee orientation. This is a great opportunity to introduce new employees to health promotion programs. Managers should work closely with those who conduct orientations to maximize new employees' initial exposure to health promotion programs. If a tour is offered, it should include a visit to the health promotion program facilities.

Risk Management and Legal Services

In some organizations, the risk management and legal functions are combined in one department; in others they are separate. In any case, the department that is responsible for approving policies, procedures, and documents is important to health promotion, and managers should develop a good working relationship with this unit.

For example, health promotion managers must obtain informed consent from participants in some programs. The legal department should review these and related documents to minimize the organization's liability. In some organizations, procedures used for health assessments and testing also must be reviewed. Because litigation in these areas can be extremely costly to the organization both financially and in terms of its public image, a good working relationship with the legal department is of utmost importance.

Sales

A relationship between the health promotion department and the sales department can be very beneficial, especially if the organization has a large sales force that is dispersed over a wide geographical area away from headquarters or other major facilities. Targeting field sales personnel with health promotion programming is very difficult.

One approach is to provide health promotion programming at sales meetings. Offerings can include fitness testing, educational programs, health risk appraisal counseling, and programs that can be completed away from health promotion facilities. For example, an employee can complete a run/walk program by keeping a log and mailing it monthly to the health promotion department. Field sales personnel would then be able to receive the same incentive rewards as participants who exercise on site.

Public Relations

A strong working relationship between health promotion and public relations can be a win-win situation. The public relations department is always looking for news items and features to improve the organization's image and exposure. The health promotion department has the same needs, and regular contact with a public relations employee can result in good publicity for health promotion. For example, if the health promotion department is planning a creative, new program to help employees lose weight, the local newspaper may want to print a feature story about it. This would provide excellent publicity for the program inside and outside the organization because it would show that management cares about its employees.

SUMMARY

A working knowledge of organizational structure is essential to the strength and stability of the health promotion program. Managers' authority is reflected by their positions on the organizational chart and the executives to whom they report. Understanding how the health promotion program fits into the organization's goals and the specific strategies used to attain those goals will help managers integrate their programs into the fabric of the workplace. Managers should strive to develop solid working relationships with other departments that capitalize on the natural synergy that may exist among any related functions.

In this chapter we discussed the relationships among departments in an organization. Each department is composed of individual employees, and the attitudes and efforts of employees affect how the organization functions and performs. In chapter 5 we consider leadership, motivation, and supervision.

KEY TERMS

authority (p. 23)
functional organizational chart (p. 24)

goals (p. 29)
line position (p. 24)
matrix organizational chart (p. 25)
power (p. 23)
project organizational chart (p. 24)
staff position (p. 24)
strategy (p. 29)

SUGGESTED RESOURCES

Anton, T. (1989). *Occupational safety and health management*. New York: McGraw-Hill.

Association of Research Libraries. (1986). *Organization charts*. Washington, DC: Office of Management Studies.

Balliet, L. (1987). *Survey of labor relations*. Washington, DC: Bureau of National Affairs.

Connor, P., & Lake, L. (1988). *Managing organizational change*. New York: Praeger.

Dessler, G. (1991). *Personnel/human resource management*. Englewood Cliffs, NJ: Prentice-Hall.

Dickman, F., Challenger, B., Emener, W., & Hutchison, W. (1988). *Employee assistance programs—A basic text*. Springfield, IL: Charles C Thomas.

Dilenschneider, R., & Forrestal, D. (1987). *The Dartnell public relations handbook*. Chicago: Dartnell.

Gomez-Mejia, L. (Ed.) (1989). *Compensation and benefits*. Washington, DC: Bureau of National Affairs.

Greiner, L., & Schein, V. (1988). *Power and organization development*. Reading, MA: Addison-Wesley.

Gutknecht, D., & Miller, J. (1986). *The organizational and human resources sourcebook*. Lanham, MD: University Press of America.

Leap, T. (1991). *Collective bargaining and labor relations*. New York: Macmillan.

Robinson, K. (1988). *A handbook of management training*. London: Kogan.

Sutton, I. (1992). *Process reliability and risk management*. New York: Van Nostrand Reinhold.

Waterman, R. (1987). *The renewal factor: How the best get and keep the competitive edge*. New York: Bantam Books.

Wilkins, A. (1989). *Developing corporate character: How to successfully change an organization without destroying it*. San Francisco: Jossey-Bass.

Chapter 5

Managing Employees

All organizations are made up of people. To manage effectively, managers must develop an understanding of people's motives and needs. In a health promotion program, this will enable managers to have an impact on participants and their work environments. It also will enable managers to supervise their employees effectively. For these reasons, program success depends in part on working with and meeting the needs of individuals.

Effective management of employees involves leadership, motivation and productivity, and supervision.

LEADERSHIP

A leader maintains the structure and goal direction that are necessary for a group to perform effectively. In a business organization, leadership responsibilities may include job design, staffing, scheduling, training, budgeting, planning, and marketing. However, it is not only skill in these areas that determines a leader's effectiveness but also leadership style. There are many styles of leadership.

Leadership Styles

No leadership style is appropriate for all situations. The effective style is the one that accurately matches the personality traits and work habits of the leader with those of the group. Because no simple system exists for determining the appropriate leadership style, health promotion managers need to understand several styles. We will describe four leadership styles. There are other styles; also, managers may use a combination of two or more styles.

Laissez-Faire

In the laissez-faire style of leadership, the manager allows the group members wide latitude to act. At first there may appear to be an absence of leadership. In reality, the manager provides some direction but gives the individuals and the group considerable freedom in decision making. Also, the manager has little involvement in performance control or appraisal.

Because the laissez-faire leadership style can result in disorganization and duplication of effort, it should be used only in situations where group members are capable of assuming much responsibility and following through on their commitments. For example, this style might be effective in a research laboratory, where scientists could be allowed to pursue their work under broad guidelines.

Autocratic

The autocratic leadership style is the opposite of the laissez-faire approach. An autocratic leader provides a high degree of direction to the group, and members have little input into the planning and control process. Performance control and appraisal are important components of this leadership style.

When a leader uses the autocratic style, group members are likely to become dissatisfied. Although management research has shown that workers like some structure, they respond negatively when the structure is overly rigid. Obviously, different types of work groups will tolerate structure differently. In general groups involved with production are more tolerant of structure than groups that perform research, sales, or technical support. On the whole, group members tend to dislike the autocratic style. However, there are situations where autocratic leadership is warranted, such as military exercises and some high-volume production operations.

Bureaucratic

A leader who uses the bureaucratic style relies heavily on the organization's rules and regulations. Both leader and group members closely follow specific procedures defined by upper management. Leaders generally do not make difficult decisions themselves but refer to policies and procedures. This may give group members the impression that the leader has no decision-making responsibility.

The bureaucratic style closely parallels the autocratic style. The major difference is the source of the direction. Autocratic leaders develop their own direction, whereas bureaucratic leaders obtain their direction from the organization's policies and procedures. In both cases the group members have little input into the decision-making process. The bureaucratic style is most effective in very large organizations such as governments where there is a need for control. However, group members may feel stifled by the weight of bureaucracy.

Democratic

The democratic leadership style allows for input from group members. In some settings the group may vote on all matters; in others the leader makes all decisions after considering input from the group. Because group members participate in the decision-making process, this style of leadership is also called participative.

Of the four leadership styles described, the one most favored by group members is the democratic style. Because group members have input into many decisions, they are more likely to be committed to achieving the group's objectives. This can result in improved productivity and greater acceptance of change. Although there are situations where democratic leadership is not appropriate, such as the military, when used appropriately it can be very effective. For example, several automobile manufacturers, particularly Japanese companies, are using this leadership style effectively in production.

The best leadership style for a group must be based on an assessment of members' personalities and working styles and the organization's business climate, including its financial position, degree of competition, and the types of products produced or services provided.

Leadership for Health Promotion

Health promotion professionals need to know about leadership for several reasons:

- To better understand their supervisors
- To manage employees effectively
- To understand the organization to which their program participants belong

The first two reasons are obvious; the third may need further explanation. To influence the health behaviors of program participants, it is helpful to understand the leadership style used in the participants' organization. This can help managers make better judgments about program design and promotion.

For example, if the leadership style is autocratic, the manager might want to promote the program to the leaders with the expectation that others will be influenced by leaders' interest in the program. However, in an organization with a democratic style of leadership it might be more effective to focus the promotion on individuals with informal power, some of whom should be in the health promotion advisory group.

In the health promotion area, effective management generally will involve some form or component of the democratic health promotion leadership style. Health promotion professionals are usually well educated and technically oriented, and they place a high value on having input in decisions. Of course, managers also must consider their supervisors' expectations. Sometimes managers must compromise between their own ideas and those of their supervisors. This may require the supervisors to use the bureaucratic or autocratic style of leadership.

There is no formula that can correctly determine the leadership style to use in health promotion or any other field. Selection of an appropriate style should be based on a careful assessment of personalities of group members and supervisors as well as the manager's own traits. Other factors to consider are the organization's structure, objectives, and competitive environment. Finally, managers should regularly evaluate and modify their leadership styles.

MOTIVATION AND PRODUCTIVITY

Motivation and productivity are of concern to health promotion professionals for several reasons. Motivation is necessary to maximize employees' output and to maximize the benefits to participants in health promotion programs. Productivity is frequently used to justify the commitment of resources to health promotion programs.

Needs Hierarchy

A pioneer in motivation theory was Abraham H. Maslow (1970). He developed a hierarchy of human needs that he termed Need Theory. Maslow identified six basic human needs, which he ranked in order of importance (see Figure 5.1): physiological, security, affiliation, esteem, integration, and self-actualization.

The lowest level of human needs are physiological: food, water, rest, shelter, and

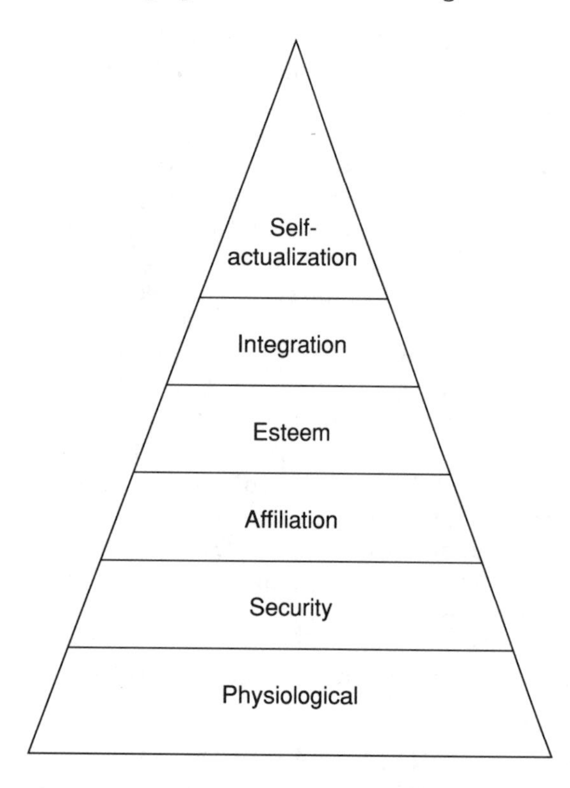

Figure 5.1 Maslow's hierarchy of human needs.

other physical necessities. In general, employees work to obtain the money to purchase the necessities for their and their families' survival.

When a person's basic physiological needs are met, other needs become important. The second level of the needs hierarchy is security, which includes protection against danger and threat. To address this need, many employers offer benefit packages of medical and life insurance that help protect their employees from the uncertainties of life. Safe working environments and job security also are related to security needs. Health promotion programs can help employees meet security needs by improving their health.

The third level of the needs hierarchy is affiliation: the need to be part of a group or to be accepted and respected by peers. Employees who are able to meet their affiliation needs are generally more productive than those who are not able to satisfy these needs. Managers must be sensitive to this need and try to help employees meet it. From the health promotion perspective, employees can satisfy affiliation needs in part by participating in exercise groups or other activities.

Closely following the need for affiliation is the need for esteem. This is the need to feel respected and worthwhile. Managers can help employees meet their esteem needs by encouraging them to develop professionally within the work group and by providing reinforcement and recognition. Even receiving a T-shirt or being featured in a health promotion newsletter can help employees meet their esteem needs.

At the fifth level of the pyramid is the need for wholeness or integration. This includes the need for knowledge, understanding, and consistency in a fair, well-organized working environment. Health promotion activities, as part of that environment, can provide knowledge of lifestyle habits and good health, leading to a feeling of wholeness.

The highest level of the needs hierarchy is self-actualization. This can best be defined as reaching one's potential. It may include the opportunity to be creative and to achieve desired objectives. Each person's criteria for self-actualization are different. Some are better able to meet this need on the job; others find more opportunities at home with their families.

Some people seek self-actualization by pursuing health and fitness, and health promotion managers can play an important role in helping employees meet those needs. If managers can create a health consciousness in the work environment, more people are likely to consider optimal health a self-actualization need and adopt healthier lifestyle habits. This clearly would benefit the people and their employers.

When people have needs, they are motivated to meet those needs. If employees' needs are congruent with those of the organization, meeting employees' needs generally will enhance productivity. This can have far-reaching benefits. By understanding the hierarchy of needs and how it relates to organizations, managers can have a greater impact on health and productivity.

Expectancy Theory

The expectancy theory developed by Vroom (1964) is based on his definition of motivation as a function of needs and the extent to which a person believes or expects that the needs will be met. This theory is shown below in equation form.

$$\text{Motivational Force} = \text{Expectancy} \times \text{Valence}$$

In this model, valence is defined as a person's positive or negative response to a specific outcome and the degree of the preference or aversion. Expectancy is defined as the belief that a specific action will be followed by a specific outcome. The greater the preference for something and the greater the expectancy that the preference can be obtained, the greater the motivation for working toward it.

The application of expectancy theory to health promotion is twofold. Health promotion professionals need to motivate their employees and the participants in their programs. To improve the motivation of participants, managers must impart the value and desirability of good heath. This will improve the valence rating.

Second, managers must help participants develop the expectation that they can achieve their health-related goals. This means that goals and expected outcomes must be realistic. Health promotion managers also must consider these factors when working with their employees. Managers must help employees develop realistic expectations of the outcomes of specific actions.

To understand employees' expectations, it is helpful to understand their motivations. For example, many health promotion professionals often reach a point where they need to have more responsibility and autonomy. Managers must become aware of these employees' expectations from their jobs and devise ways to help them develop professionally by increasing their expectancy that continued hard work will benefit them in salary increases and promotions. This technique can improve performance in the health promotion department and enhance the organization's public image.

Job Satisfaction

Most people seek satisfaction in their jobs. Effective managers know the components of satisfaction and dissatisfaction and use this knowledge to make personnel decisions that maximize their employees' job satisfaction. Due to conflicting research there is considerable controversy about what constitutes job satisfaction, yet the motivation-hygiene theory developed by Herzberg, Mausner, and Snyderman (1959) provides useful insights into the job satisfaction problem. The research underlying this theory was conducted on salaried workers, specifically engineers and accountants. Because many health promotion professionals can be considered white collar, this is a reasonable model to use with this population.

According to this theory, motivators are factors that increase job satisfaction and provide motivation for superior performance. Motivators include achievement, recognition, the work itself, responsibility, and advancement. The absence of motivators, however, does not lead to dissatisfaction.

Hygiene factors do not result in job satisfaction when present but when perceived to be inadequate create dissatisfaction. Factors that may be perceived negatively, such as conflicting organization policy and administration, inappropriate supervision, low salary, poor interpersonal relations, and poor working conditions, can result in job dissatisfaction. Most of these factors relate either directly or indirectly to the organization's working environment.

A major criticism of this theory is the identification of salary as a hygiene factor. Salary has been identified as a motivator in other theories and research. To be safe it would be advisable to consider the possibility that salary is a motivator and not a hygiene factor when making personnel decisions.

Job Performance

Many studies have considered the relationship between job satisfaction and job performance. The results of these studies are mixed and inconclusive. In general, job performance depends on more than job satisfaction. Although job satisfaction warrants consideration, other variables also are important, such as feelings of accomplishment, pay, promotion, and status.

Another factor that affects job performance and productivity is the work group's attitude. A cohesive group tends to be more productive if management's expectations are realistic. If management has unrealistic expectations, a cohesive group is more likely to combine forces against management and be less productive. For this reason, it is necessary to develop cohesive work groups and manage them effectively.

One of the most significant factors in productivity often is technology. State-of-the-art computers and software can measurably improve the productivity of the health promotion department. (Computer technology is discussed in more detail in chapter 10.) Fitness testing equipment and efficient communication systems also can save time and money. If technology is managed appropriately, its impact can be greater than that of the social factors of management.

Managers should take all possible measures to improve employee performance and productivity. Managers of health promotion must consider job satisfaction, group attitudes and cohesiveness, and technology. By maximizing these factors, the health promotion department can improve the quality and cost effectiveness of its services and more forcefully justify its programs to management.

SUPERVISION

Although managers are responsible for many different types of resources, for many the most important resource is people. This is particularly true in the health promotion field. Except for the smallest programs, the manager of health promotion supervises at least one employee. The manager is also responsible for supervising many other individuals while they are participating in health promotion programs. For this reason, a basic understanding of supervision is necessary for all health promotion professionals.

Challenges

Numerous challenges confront supervisors in the health promotion area. These include education levels, working conditions, technology, attitudes, and decision-making processes.

Education

Whereas in some industries supervisors face the problem of dealing with unskilled workers, in health promotion the challenge is to find individuals with the diverse skills required for success in the field. Because this is a relatively young and growing field, it is difficult to find consensus on the most desirable educational background. Areas frequently considered are health education, business, counseling, exercise science, and communication. Health promotion managers must evaluate their needs carefully and use an effective job design process (discussed in chapter 6).

Working Conditions

The work environment is another challenge of supervisory management. Working conditions go beyond safe and aesthetic surroundings. Other issues are working hours, dress codes, policies and procedures, and job security.

In the 1960s there was considerable discussion about a 35-hour work week and increased vacation time, and predictions that these would become a reality in the 1980s. In most organizations these expectations never materialized. However, some organizations operate on a 4-day week, and others use a flex-time system under which employees must work a full week but have some choice about arrival and departure times.

Of great concern to health promotion professionals is job security. When promoting employee health was a new concept, many organizations implemented programs without adequate support. If financial problems later arose, the programs often were reduced or eliminated. As the health promotion field matures, this situation appears to be improving. Nevertheless, managers must be aware of this potential problem and try to protect employees' jobs by positioning health promotion more favorably in the organization.

Technological Change

For managers the key to success in dealing with technological change is continuing education. Managers must stay abreast of changes in technology and prepare their employees to respond constructively to change. As mentioned earlier, access to and familiarity with modern automation systems will be vital to the success of many health promotion functions. For this reason, managers should pursue appropriate computer training on an ongoing basis.

Attitudes and Behavior Toward Authority

Another challenge that confronts managers is changing attitudes in society and in the workplace. In the 1960s there were radical changes in values and challenges to authority in government, business, education, religion, and the family. However, in the 1970s and 1980s there was a gradual shift back to more traditional attitudes and behaviors. Managers must be aware of changing social attitudes and prepare themselves and their employees to deal constructively with these changes.

In addition to social attitudes, health promotion managers must be sensitive to attitudes in the organization and to the authority structure. There are two reasons for this. First, health promotion is not a traditional business discipline like marketing, accounting, and production management that have well-established roles in the organization. Second, health promotion frequently does not directly affect the organization's profit. Managers must try to demonstrate the value of the health promotion department and present themselves as future leaders who are comfortable in a boardroom setting.

Participation in Decision Making

One result of the changing attitudes toward authority is the desire of many people to have a greater voice in the decision-making processes. Health promotion managers must learn to welcome employee participation in decision making. It is possible to structure job responsibilities to allow more employee input while still meeting departmental and organizational objectives.

Managerial Functions

Traditionally, supervisory management has been the link between rank-and-file employees and senior management. Supervisors must have technical competence to accomplish assigned tasks. They also must develop the skills they will need to execute management responsibilities.

Management Skills

Health promotion managers must acquire and refine management skills, such as planning, organizing, staffing, directing, and controlling. Many health promotion professionals fail to develop these skills because they are focusing their efforts on technical skills. Although technical skills are essential, a technically perfect health promotion program will fail if it is not skillfully managed.

As with any skill, it takes time, practice, and determination to develop and refine management skills. Health promotion managers should attend management workshops and seminars, work with other managers in the organization to develop new skills, and apply management principles on the job.

Technical Skills

Health promotion may encompass such varied disciplines as nutrition, exercise, stress management, and smoking cessation. By understanding each of these areas, managers will be better able to hire qualified professionals and help them perform effectively.

Management Responsibilities

The supervisor's management responsibilities are planning, organizing, staffing, directing, and controlling. These functions will be discussed in detail in later chapters.

Working Styles

Every employee has a unique personality and consequently a unique working style. By developing an understanding of human personality, supervisors can learn to manage in a way that takes account of the differences among individual employees.

Understanding Human Personality

People's personalities are influenced by heredity, gender, family environment, race, socioeconomic status, and religious and cultural values.

Managers should be aware of the factors that may have influenced the personality development of employees. Managers can use this knowledge to make decisions about how to supervise each employee.

Supervision of Different Working Styles

As pointed out earlier, working style tends to be a function of personality. Some employees like direction, whereas others prefer autonomy. Some like to work in groups; others prefer to work alone. Many other preferences can be identified. Effective supervisors learn their employees' personalities and working styles and maximize their productivity by assigning specific tasks to the appropriate employees and by using communication techniques tailored to specific employees' styles.

Authority and Delegation

Managers are given specific authority to manage their departments or functions. In many situations it is desirable to delegate some authority to another employee. Because the decision to delegate can have a strong impact on the overall success of the department or work group, managers must make careful decisions about what authority to delegate and to whom.

Authority

The concept of authority is discussed in chapter 4. It is defined as a person's right to provide positive or negative input to the behavior of others. The source of a manager's authority is the immediate supervisor. However, this is formal authority. Frequently, when working with employees, formal authority alone is not adequate. The authority derived from one's ability to relate to others is also important, and in some situations it can be the determining factor in the quality of a department's work.

Delegation of Authority

For managers to be effective leaders, they must delegate some authority. Delegating authority is defined as assigning decision-making to employees so they can work with specific constraints. By delegating work appropriately, managers can maximize the department's productivity.

Delegation

There are two types of delegation. Downward delegation of responsibility to subordinates is more common. However, there are situations where it is appropriate to delegate responsibility upward to a superior.

Downward delegation involves three steps. First, managers must decide which duties can be delegated and to whom. Second, managers must determine how much authority to grant. The employees to whom the duties are delegated must know to what extent they can make commitments, use resources, and make related decisions. The third step is to establish responsibility on the part of the employee to accomplish the assigned duties.

All three steps are required for the delegation process to be successful. Also, if it becomes necessary to change one of the steps, the other two also will have to be modified. For example, if it becomes necessary to increase the amount of authority delegated, then the person to whom it is delegated may need to change because the initial designee is not qualified to handle the additional responsibility. Additionally, the new designee must accept the responsibility.

Upward delegation involves the same basic process as downward delegation except there is not a real transfer of authority. An example of upward delegation is asking a superior to write a letter of support for a proposal to senior management. Although the manager of the affected department could write the letter, it might have greater impact if it came from the manager's superior. Obviously, the superior must accept the obligation for the transfer to be complete.

Performance Management

Another supervisory responsibility is the management of employee performance. Most organizations have a formal procedure for appraising performance. This subject is discussed in detail in chapter 8. In this section we address day-to-day performance issues.

Monitoring Performance

Managers typically monitor employee performance in two ways. First, managers directly observe employees. This is the most reliable source of information about performance. This method is very time consuming, and the time invested must be weighed against the value of the information obtained. Nevertheless, employee observation is an important responsibility, and every manager should allocate some time to this function.

Second, managers can evaluate written reports prepared by others. The reports should contain accurate, clear, and complete summary information. Managers should be able to read the reports quickly and identify any areas where performance is above or below standard. Follow-up action then can be directed to these areas.

Documentation

An important part of performance management is documentation. Written records are frequently needed for reference and verification. When an incident occurs that deviates from standard practice, a written report should be made. Deviations can range from tardiness to serious errors in judgment or performance.

Formal documentation can be done on a standard form or in a letter to the employee. Documentation should include

- the employee's name,
- details of the deviation from standards,
- the date and time of the deviation,
- future expectations, and
- the consequences of failure to improve.

A copy of this document should be given to the employee, and a copy should be placed in the employee's file. Should it become necessary to terminate the employee, such documentation can be used to support the decision.

In some situations, informal documentation can be used. Managers can make notes about an employee's behavior, such as failure to complete a task on time. If the problem persists, these notes can be the basis for formal documentation. If the incident later is seen to be isolated, the notes can be discarded.

Formal documentation is essential for legal reasons. Federal and state laws protect employees from many forms of discrimination. If a discharged employee sues the organization for discrimination, the supervisor and others in management will have to support their assertion that dismissal was for cause and legal. For this reason, written documentation of the events that led to the termination is extremely important.

TOTAL QUALITY MANAGEMENT

Due to the increase in international competition and the increased focus on quality goods and services, many organizations are using total quality management (TQM), an integrated management system with the goal of customer satisfaction. It is founded on three principles of action (Hunt, 1993):

- To focus on customer satisfaction
- To continuously seek improvements
- To involve the entire work force

Customer satisfaction and seeking improvements are addressed in the chapters on marketing. For TQM to be successful, employees must be empowered, and there must be a focus on teamwork.

One premise of TQM is that senior managers are ultimately responsible for quality. The decisions they make affect customers, employees, suppliers, and shareholders. To improve quality, managers make decisions with the following considerations:

- Senior management must lead by example.
- Training and development must be supported.
- A support structure of employees must be available to senior management to implement change.
- Rewards and recognition must be given to those who excel in the quality process.

- Data must be collected and used.

TQM techniques can be used in the health promotion department. For example, an employee advisory committee can be formed to represent program participants. The group can provide ideas and insights for program operations. Creating such a group can empower employees and improve the quality of the program.

The specifics of TQM are beyond the scope of this text. However, health promotion professionals should become familiar with the concepts in order to work effectively with organizations that use this approach.

SUMMARY

An organization's most valuable assets are its employees. This is particularly true in the health promotion area because it is very labor intensive. Managers must develop a good understanding of the needs and motives of their subordinates and of health promotion program participants.

A manager should be able to

- understand different leadership styles and know when each is appropriate,
- apply appropriate theories to motivation and performance,
- effectively use authority and delegate responsibility,
- manage performance, and
- document performance.

As we have seen in this chapter, an understanding of individuals' personalities, needs, and work styles is essential to the organization's success. In the next chapter we explain how to attract, hire, train, and develop high-caliber employees.

KEY TERMS

affiliation (p. 40)
autocratic leadership (p. 38)
bureaucratic leadership (p. 38)
delegation of authority (p. 45)
expectancy (p. 41)
integration (p. 40)
laissez-faire leadership (p. 38)
self-actualization (p. 40)
total quality management (p. 46)
valence (p. 41)

SUGGESTED RESOURCES

Brion, J. (1989). *Organizational leadership of human resources: The knowledge and the skills*. Greenwich, CT: Jai Press.

Conger, J., & Kanungo, R. (1988). *Charismatic leadership: The elusive factor in organizational effectiveness*. San Francisco: Jossey-Bass.

Fiedler, F. (1987). *New approaches to effective leadership: Cognitive resources and organizational performance*. New York: Wiley.

Hawkins, P. (1989). *Supervision in the helping professions: An individual, group, and organizational approach*. Philadelphia: Open University Press.

Hunt, V. (1993). *Managing for quality: Integrating quality and business strategy*. Homewood, IL: Business One Irwin.

Sashkin, M., & Kiser, K. (1991). *Total quality management*. Seabrook, MD: Ducochon Press.

Schein, E. (1985). *Organizational culture and leadership*. San Francisco: Jossey-Bass.

Chapter 6

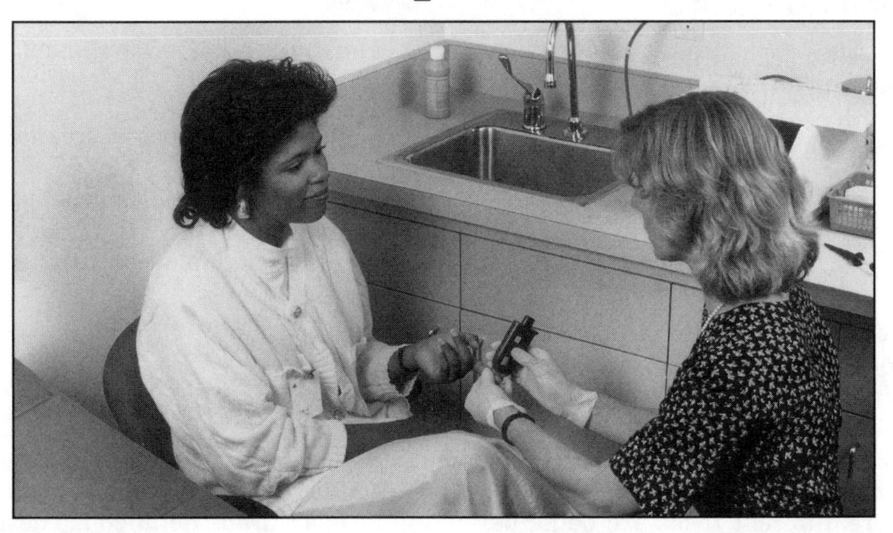

Job Design and Evaluation

An important part of most managers' work is to design jobs for the employees they supervise. In health promotion organizations, this function is likely to be handled by senior management. In a large organization that has a health promotion department to serve employees, it is likely to be performed by the health promotion manager in conjunction with human resources professionals. In either case it is important for health promotion professionals to have a basic understanding of the job design function.

In a health promotion setting, the first step in the job design process is to develop a program plan. The next step is to determine what functions must be performed to deliver each of the services the department will offer. This is called a job analysis. From this information job descriptions can be written that detail the responsibilities of each position. The next step is to develop job specifications: the skills and educational background needed to satisfactorily perform the job. Finally, performance standards should be developed for each position. Each of these steps is important for successful job design, and a discussion of each follows.

PROGRAM PLANNING

The planning process is discussed in detail in Part V. The steps outlined there should be followed when designing jobs. Some important considerations for program planning and job design follow.

Determination of Program Objectives

Before a position can be designed, the manager must clearly define the program objectives. Health promotion professionals must be sure that the program objectives are consistent with the organization's mission, goals, and objectives (pp. 201-202).

Program Planning and Job Design

Once the program's objectives have been clearly defined, the structure of the program (or department) can be planned. This process involves dividing the total task into manageable components (see Figure 6.1). Work should be subdivided horizontally where different tasks are being performed by different people and vertically where higher-level employees are responsible for supervising more people, coordinating more subgroups, and planning more complex operations.

Job Enrichment and Enlargement

In chapter 5 we discussed self-actualization and the importance of job responsibilities in satisfying this need. This should be a significant consideration in job design. In general, employees are happier and more motivated if their positions provide responsibility, challenge, and a degree of autonomy.

There are several ways to enrich or enlarge jobs. Job enrichment involves increasing management responsibilities while decreasing responsibilities for day-to-day functions. Job enlargement can be vertical or horizontal. Vertical job enlargement strategies include

- assigning employees more planning functions,
- assigning employees more controlling functions, and
- allowing employees more team participation.

Horizontal job enlargement strategies include

- giving employees a greater variety of tasks,
- giving employees a greater number of tasks, and
- rotating employees through different jobs.

JOB ANALYSIS

The purpose of conducting a job analysis is to obtain the information necessary to write a new, accurate, detailed job description or to update an old one. In larger organizations, job descriptions may be developed by the human resources department or in cooperation with it. Otherwise, health promotion managers can perform a job analysis using one or more of the following techniques:

- Observations of employees doing similar jobs
- Interviews of managers and employees in similar positions
- Questionnaires given to managers in similar positions in other organizations
- Evaluations of behaviors that characterize very good or very poor performance

Health promotion operations often have only a few employees. In this situation, it may not be practical to use interviews and observations of employees in similar positions because these may not exist. Instead managers can use benchmarking techniques, such as contacting other professionals in the field and using interviews and/or questionnaires to obtain information from them. Observations of employees working

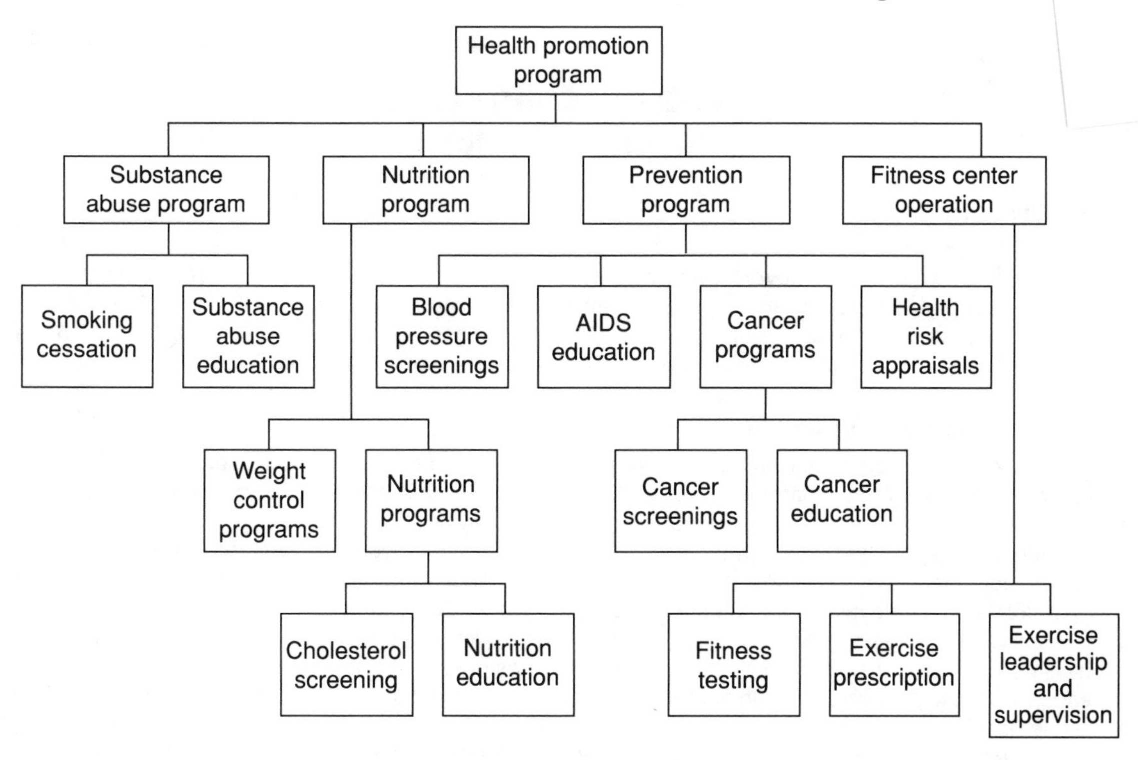

Figure 6.1 The first step in program planning is to divide the program into manageable and efficient units.

can be made at other organizations to obtain a list of functions they perform.

When conducting a job analysis, managers need to obtain several kinds of information. These might include

- what activities are performed;
- how activities are performed;
- when activities are performed;
- why activities are performed;
- what equipment is used;
- what interactions with others are required;
- what the working conditions are; and
- what training, skills, and abilities are required.

The information obtained in a job analysis, along with data from applications, interviews, and tests, can be used to develop accurate job descriptions that will help the manager build a skilled, efficient staff.

JOB DESCRIPTION

The next step is to develop a job description that provides specific information about the position. The first step in developing a job description as shown in Figure 6.2 is to conduct a job analysis, as outlined earlier in the chapter. Sometimes a job description must be written for a new position. In other situations it is necessary to revise job descriptions for existing staff members who will be assuming additional or different responsibilities. Any new responsibilities should be added to the job description.

JOB DESCRIPTION
Program Director

Job Summary

The program director is responsible for overseeing all health promotion activities; supervises staff; ensures coordination of the development and implementation of fitness programs, health education activities, program promotion, and special events; ensures proper maintenance of participant records; reports to management; ensures maintenance of equipment and facility; ensures safe, clean environment for the center; carries out other projects and duties that contribute to the program.

Job Duties and Responsibilities

Oversees all fitness center and health promotion activities: Ensures that participants' needs are met satisfactorily; verifies that fitness center activities comply with established company standards; and coordinates daily, monthly, and quarterly reporting activities.

Supervises staff: Hires, trains, and discharges personnel as necessary; delegates work assignments; approves staff schedules; evaluates staff performance; and ensures staff development.

Coordinates the development, integration, and implementation of fitness programs, health education activities, program promotion, and special events; devises periodic programs to encourage facility use and stimulates interest in the program; creates graphic displays and other props as health education tools; prepares periodic newsletters to update participants on fitness center programs and timely health information; plans and oversees special health and fitness events.

Ensures proper maintenance of participant records: Maintains records of participants' (1) attendance at fitness facility, (2) exercise frequency, (3) changes in physiological conditioning, (4) injury or illness incidents at the facility, and (5) membership payroll deductions or fee payments. Retests participants and updates exercise program; ensures follow-up with participants with low facility attendance; oversees all computer operations; and verifies records for accuracy.

Interfaces with company management and peripheral departments: Maintains ongoing communication with company management, medical department, and other functional areas as appropriate; prepares annual budget; develops annual operating plan; handles inquiries by media and other interested parties; refers callers, when applicable, to appropriate managers.

Conducts fitness testing: Evaluates cardiorespiratory and muscle condition of program participants; prescribes exercise programs; and instructs participants on proper techniques for carrying out recommended exercise program.

Ensures safe and clean environment at the center: Oversees maintenance and janitorial contracts; ensures equipment is well maintained and operating properly.

Job Specifications

Graduate degree in physical education, exercise physiology, kinesiology, or health education from an accredited college or university; prior experience in program management and staff supervision; certificate in CPR and first aid mandatory; communication skills to instruct and interact effectively in a corporate setting also required.

Figure 6.2 Clear job descriptions are essential in defining employees' responsibilities and in evaluating their performance.

Uses of Job Descriptions

A well-written job description can be useful to a health promotion manager in several ways, including the following:

- Using preliminary drafts to generate group discussions about the position and how it relates to other positions in the organization/department
- Developing job specifications or job requirements
- Developing performance standards
- Planning for work force needs and recruiting
- Screening resumes
- Structuring interviews
- Orienting new employees to the responsibilities and duties of their positions
- Conducting evaluations to determine the relative worth of jobs in an organization

Health promotion managers can use job descriptions for most of these purposes. Worksite health promotion managers may not be involved in planning work force needs and conducting job evaluations. These are typically human resources functions. However, the manager of a large health promotion organization may have these responsibilities. Also, managers of worksite programs must understand these functions and cooperate with the human resources department.

By establishing health promotion positions as having high relative worth in the organization and integrating health promotion staffing needs into work force planning and recruiting, managers can increase the perceived value of health promotion to the organization. This can improve the department's standing in the organization and make it easier to obtain resources.

Developing the Job Description

Much of the responsibility for developing job descriptions belongs to the human resources department. However, when professionals lack the expertise to write job descriptions for certain positions, they work with the managers of those departments to develop descriptions. The human resources professionals will provide appropriate guidance, but health promotion managers must supply the specialized knowledge to make the job description accurate and useful.

Typical job description items for a health promotion specialist are

- orients new members to the health promotion program,
- conducts fitness assessments,
- develops timely health promotion programs,
- publishes the health promotion newsletter,
- represents the health promotion department on the public relations committee,
- evaluates health promotion program effectiveness,
- plans heart-healthy meals in the cafeteria, and
- maintains participant records.

Accuracy of the Job Description

For job descriptions to be useful, they must be accurate. Department managers need accurate job descriptions to develop an efficient, productive staff. The human resources department uses job descriptions for job evaluation and to determine compensation. Because people in this and other departments often are not familiar with the functions and responsibilities of health promotion professionals, they rely heavily on job descriptions for information. Accurate job descriptions provide a realistic picture of duties, training needs, and performance standards.

JOB SPECIFICATIONS

After the job description has been written, the information it contains can be used to develop specifications for the job. Job specifications translate the responsibilities of the job into the skills and knowledge required to perform the job, such as education level or basic life support certification. (See job specifications in Figure 6.2.)

Purposes of Job Specifications

Job specifications identify the qualifications required to successfully perform the duties outlined in the job description. The specifications usually are appended to the job description. The specifications can be used in hiring and for job evaluation. For example, the information in the job specifications can help managers determine the questions on which to focus in the job application, appropriate questions to ask in the interview, what types of tests, if any, to administer to the applicant, and which references to check, including what questions to ask the people used for references. Job specifications are also used to determine the value of a job to the organization and to make compensation decisions.

Determining Job Specifications

Job specifications usually are determined by the department manager and human resources professionals. When job specifications are being written for health promotion positions, at least one professional from that department should be involved.

In developing job specifications, the managers should consider their experiences, the job description, and the organization's future plans.

In organizations that have had little experience in the area of health promotion, it may be necessary to consult with other professionals in the field. This may be done when conducting a job analysis in a different organization. The organizational planning process and job description information should be readily available.

Job specifications must comply with government regulations that are intended to prevent the wrongful exclusion of qualified applicants. Human resources employees should handle the legal aspects of developing job specifications.

Evaluating Job Specifications

The purpose of developing job specifications is to ensure that the people who are hired have the education, skills, and experience to perform their duties competently. Periodically reevaluating job specifications in terms of the tasks that must be performed to deliver program services will enable managers to determine whether any employees are under- or overqualified. Managers can revise job specifications as appropriate to meet the needs of the health promotion program.

PERFORMANCE STANDARDS

After job specifications have been written, managers must develop standards for measuring performance. Examples include: conducting two valid fitness tests per hour, publishing the newsletter with no typographical errors, or implementing a marketing plan that results in 30 new members per month.

Purposes of Performance Standards

Performance standards are quantitative and qualitative criteria by which managers assess employees' execution of the responsibilities specified in their job descriptions. Performance standards should specify how much work is to be accomplished, within what amount of time, and what constitutes acceptable performance. In chapter 23 we discuss work plans and develop these concepts further.

Performance standards also can be used as guidelines to help employees develop individual goals that are consistent with departmental and organizational goals. Employees who are permitted to set objectives that are meaningful to them will be more committed to the organization and more motivated to perform at an acceptable level and qualify for promotion.

Development of Performance Standards

Performance standards usually are separate statements that supplement job descriptions and are part of the work plan development process. Standards can be included in job descriptions. However, structuring them as supplements gives managers the flexibility to revise standards as circumstances change, without having to rewrite the job description.

SUMMARY

Whether managers are creating new positions or redefining existing positions, they should follow these job design steps:

- Determine objectives.
- Conduct a job analysis.
- Develop the job description.
- Establish job specifications.
- Develop performance standards.

By taking an organized approach to job design, managers will be better able to justify health promotion positions to senior management. Carefully prepared job descriptions, job specifications, and performance standards also will facilitate the staffing functions discussed in the next chapter.

KEY TERMS

job analysis (p. 50)
job description (p. 51)
job design (p. 50)
job specifications (p. 54)
performance standards (p. 54)

SUGGESTED RESOURCES

Kaufman, L. (1986). *Job evaluation systems: Concepts and issues*. Kingston, ON: Industrial Relations Centre, Queen's University at Kingston.

Manese, W. (1988). *Occupational job evaluation: A research-based approach to job classification*. New York: Quorum Books.

Steele, F. (1986). *Making and managing high-quality workplaces: An organization ecology*. New York: Teachers College Press.

Thompson, P., & McHugh, D. (1990). *Work organizations: A critical introduction*. Basingstoke, England: Macmillan.

Chapter 7

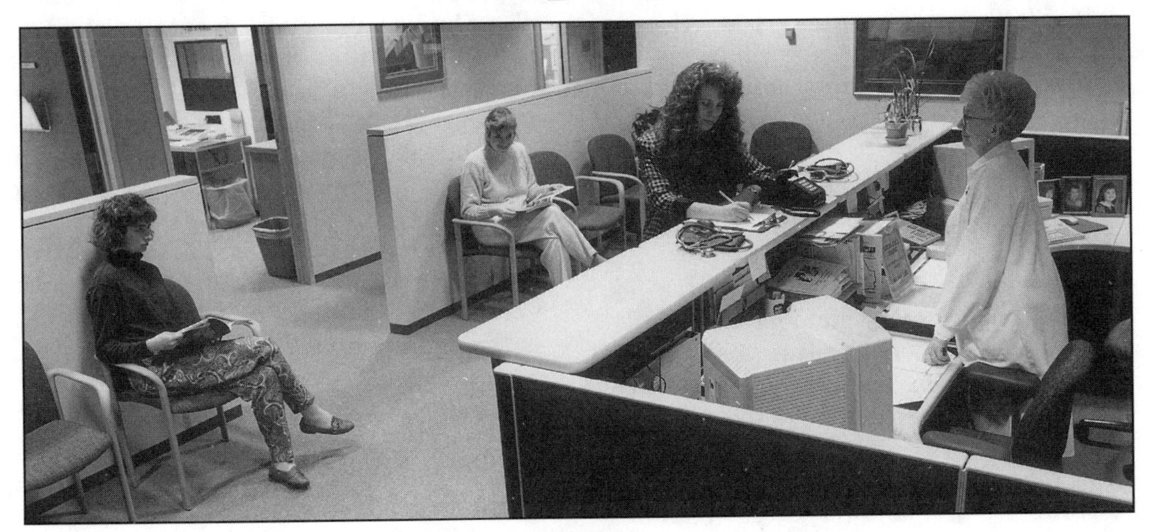

Staffing

The staffing function of an organization involves many processes that are intended to ensure that all positions are filled in a timely and appropriate manner. To manage the staffing function, it is necessary to understand processes such as

- determining personnel requirements,
- obtaining authorization to fill positions,
- developing a pool of applicants,
- evaluating applicants,
- making hiring decisions,
- inducting and orienting new employees, and
- coordinating staffing changes.

Health promotion managers may be involved to some degree with the staffing function. Although the human resources department usually provides assistance in this process, many of the fundamental responsibilities will be handled by the health promotion manager.

PLANNING AND RECRUITMENT

To minimize the amount of time positions are left unfilled, managers must develop and maintain a recruiting plan. One purpose of the plan is to project personnel needs that may be created by expansion, downsizing, or changes in program functions. A recruiting plan also can be used to create a network for identifying potential applicants when they are needed, especially on short notice. The network could consist of other health promotion managers and university professors who know of qualified people who might be interested in a job. This will decrease the time required to develop an applicant pool. When a position opens, the manager can obtain authorization to fill the position and develop an applicant pool quickly.

Personnel Planning

In most organizations, responsibility for staffing belongs to the human resources department. Planners analyze skill levels, current and expected position vacancies, and current and expected expansions or reductions. Then they develop plans to deal with these changes.

However, human resources planning usually is involved with operations of the organization that generate income. In organizations where health promotion is an employee benefit rather than an income-generating function, health promotion managers must develop their own plans to fill vacant positions in a timely manner. Although this process is simple because there are relatively few employees in most health promotion departments, it is very important because one vacant position in a small group can put a heavy burden on the remaining employees. In planning to meet personnel needs, managers should consult with the human resources department to discuss and review plans.

Intelligent personnel planning decisions are based on accurate information, some of which can be obtained from other departments in the organization. Managers can consult human resources planners to determine if there are any expected shifts in staff size or location of employees. For example, a decision to transfer a large group of employees from one facility to another might indicate the need to move a health promotion specialist to the expanded facility.

Another resource is the accounting and finance department, which has information about budget allocations for staffing and financial projections that will affect the organization in the future. If a health promotion manager needs to increase staff to meet the demands of clients or employees, financial information may help predict the likelihood of obtaining authorization to add positions.

The next step is to obtain information about the health promotion department's employees and programs. Managers should gather information such as

- individual career plans,
- expected vacancies as a result of resignations, promotions, and terminations, and
- skills needed for current or planned program offerings.

Another major factor in personnel planning is the external environment. Among the variables to consider are the quantity and quality of health promotion professionals in the job market, as well as changes in the demand for certain educational backgrounds, skills, or experience. The availability of personnel also is affected by economic conditions. For example, in a recession, health promotion could become a lower priority, which would cause an increase in the number of health promotion professionals looking for jobs.

Obtaining Authorization to Fill a Position

Most organizations have a set procedure for filling a vacant position. Normally, the manager must complete an employee requisition form (see Figure 7.1). The form should contain the job description, job specifications, and salary range. If senior management approves the request, an applicant

Requisition to Fill an Open Position

Department ______________________________ Date ____________

Department code ______________________________

Job title ______________________________ Job title code ____________

Pay grade ____________ Salary range ____________ to ____________ Class ____________

New position? _____ yes (New position must be classified)

_____ no (Person to be replaced: ______________________________)

Date position is expected to be vacated ____________ filled ____________

Is the position to be filled by departmental promotion? _____ Organization-wide promotion? _____

Check the type of appointment:

_____ Full-time _____ Part-time _____ Intermittent _____ Seasonal

_____ % FTE _____ Temporary from ____________ to ____________

Number of hours per week __________ Shift: 1st 2nd 3rd

Workdays: M T W T F S S Work hours: __________ to __________

Work location ______________________________

Contact person ______________________________

Major duties and responsibilities (attach separate sheet for additional information):

Minimum qualifications:

Funding contact person ______________________________ Phone ____________

Present funding:

Account code	Annual amount
____________	______
____________	______
____________	______
____________	______
Total	______

Proposed funding:

Account code	Annual amount	Start date	Stop date
____________	______	______	______
____________	______	______	______
____________	______	______	______
____________	______	______	______
Total	______		

Approvals	Approved	Disapproved	Date
Department manager			
Vice president			

For Human Resource use only

Compensation

Approved class title ______________________________ Range/grade ____________

Approved by ______________________________ Date ____________
(Analyst)

Position filled by ______________________________ Start date ____________

Figure 7.1 An employee requisition form is used to obtain authorization from senior management and ensure that funds are available for the position.

pool can be developed or the selection process can begin if an adequate pool is available.

Developing the Applicant Pool

Responsibility for developing a pool of applicants usually rests with the human resources department. The role of the health promotion manager will depend on the nature of the organization. If health promotion is an employee benefit, the human resources department may need substantial input from health promotion professionals. In a large health promotion management organization, human resources specialists will be familiar with staffing requirements. In either case, health promotion professionals should know how to develop a pool of applicants on their own.

There are several methods of developing applicant pools. Common practices are summarized in Table 7.1.

Table 7.1
Methods of Developing an Applicant Pool

Method	Target
Newspaper advertisement	Entry-level positions Metropolitan areas
Journal advertisement	Upper-level positions Small communities
Job postings	Qualified people in a large organization
Placement agencies	High-level positions
College campus visits	Entry-level positions
Field recruiting	Large metropolitan areas
Maintaining files	All groups

THE SELECTION PROCESS

The process of selecting new employees has several components: a written application, interviews, the demonstration of skills (such as leading an exercise class), psychological tests, a physical exam, a drug test, and a job offer. Throughout this process it is essential to comply with Equal Employment Opportunity Commission (EEOC) and affirmative action guidelines.

The Application

Most organizations have a standard application form that is developed, distributed, reviewed, and filed by the human resources department. The application form is used to obtain information that can be used to evaluate the applicant's qualifications. An example of an application form is shown in Figure 7.2. Other functions of the application form are to disseminate information to managers who may be interested in the applicant and to store information for future use.

In designing an application form, questions must be structured to elicit information that will be reliably and consistently predictive of success in the organization. The application also must conform to all laws, including EEOC regulations. Human resources professionals are responsible for adherence to these requirements and should be consulted when questions arise.

In some cases job seekers visit a prospective employer and complete an application. In many instances, however, an applicant makes the initial contact by mail and sends a resume. In the health promotion field, the resume is a very important job-seeking tool, and managers must learn to accurately evaluate resumes. Consider the following points:

- Work experience in people-service positions: health promotion, customer service, other public-contact positions

EMPLOYMENT APPLICATION An Equal Opportunity Employer

For your application to be properly evaluated, you must answer each of the following questions as carefully and as completely as possible. Please use black ink (or type). If more space is needed for your answers, you should attach a separate sheet. Please add any additional information that will help us to evaluate your qualifications.

This application will be kept on file for six months. Date of application ____________

PERSONAL INFORMATION		
	Full name (first) (middle) (last)	
	Name you prefer to be called	Are you known to schools or prior employers by another name? If yes, by what name?
	Address (P.O. Box or street and apartment number)	
	City State	ZIP code
	Social security number	If under 18 years old, give date of birth / /
	Area code & telephone number ( )	Time to call
	Alternate telephone number ( )	Time to call
	Are you a citizen of the United States? ❑ Yes ❑ No	
	Alien registration card # ____________	Visa status

POSITION INFORMATION			
	Type of employment desired ❑ Full time ❑ Part time ❑ Summer		
	Position applied for		Years of related experience
	Job no.	Salary desired	Date available to work
	Who referred you to our company?		
	Have you ever been employed by our company? ❑ Yes ❑ No If yes, state the branch, the location, and employment dates below:		
	List relatives employed by our company and state your relationship.		

(continued)

Figure 7.2 A job application should contain personal, employment, education, and skills information.

Please indicate no fewer than your last three employers. In the space provided, list your most recent employer and position first and all positions held with that company including initial position. Attach a separate sheet if necessary. If you would like to attach a résumé, you may do so. Résumé attached ❑ Yes ❑ No

EMPLOYMENT INFORMATION

Company name	Address	City	State	ZIP code

Position title	From	To	Ending salary	Briefly describe job duties

Reason for leaving	Last supervisor's name and phone #

Company name	Address	City	State	ZIP code

Position title	From	To	Ending salary	Briefly describe job duties

Reason for leaving	Last supervisor's name and phone #

Company name	Address	City	State	ZIP code

Position title	From	To	Ending salary	Briefly describe job duties

Reason for leaving	Last supervisor's name and phone #

May we call your present employer? ❑ Yes ❑ No
Are you under any contract or restriction with a former employer? ❑ Yes ❑ No
If yes, which employer?

XYZ is an equal opportunity employer and does not discriminate in employment on the basis of race, sex, religion, national origin, disability or veteran status, or in violation of applicable federal and state statutes regarding age. The federal government requires XYZ to maintain records concerning job applicants. We would appreciate your completing all items in this section. However, you need not answer any questions in this section to which you object. This portion of your application will be detached and will be used only to prepare data on our applicants. It will not be seen by the hiring supervisor. Your cooperation is appreciated.

EEO DATA

Name		Social security number
Position applied for	Job no.	Today's date

Race or ethnic background ❑ Black ❑ Asian or Polynesian ❑ American Indian ❑ Hispanic ❑ White or other

Sex ❑ Male ❑ Female	Date of birth

Figure 7.2 *(continued)*

U.S. MILITARY SERVICE		
Branch of service	Date entered	Date discharged
Military experience relating to position applied for		
Active reservist ❑ Yes ❑ No		
Are you a veteran? ❑ No ❑ Yes If so, what dates did you serve?		
Are you a Vietnam era veteran? (Served more than 180 days of active duty 8-5-64 through 5-7-75) ❑ Yes ❑ No Disabled ❑ Yes ❑ No If yes, % disabled______		
Do you have any disabilities that would prevent you from performing the position for which you are applying or that would require special accommodations? No ______ Yes ______ Explain ______		

EDUCATION & SKILLS INFO

School name City and state	Major subject & no. of hours	Total no. of hours credit	Date attended From	Date attended To	Diploma or degree	When you will graduate
High school						
College						
College						
Graduate school						
Other (business, service, trade, correspondence, etc.)						

High school grade ___ Average ___	Cummulative G.P.A. (college) average ___out of possible ___		
Approx. (college) standing in class: ______ Out of total class of: ______			
What foreign language do you	Read	Write	Speak
Typing wpm	Dictation wpm	Machines or equipment operated	
List special awards, honorary organizations, offices held, and other activities (you may exclude those that may reveal race, religion, physical disability, martial status, or ancestry):			
List professional societies, memberships, and offices held:			
List professional licenses and state where held:			

Figure 7.2 *(continued)*

Whom can we contact in the event of emergency?
Name Address Telephone no.

Have you been convicted of a felony within the past five years? ❑ Yes ❑ No
If yes, detail briefly including date and location.

Note: A conviction will not necessarily bar an applicant from employment. Conviction records ordered sealed or expunged by a court need not be disclosed.

Please read the statements below carefully. Your signature will indicate that you understand and agree with these statements:

I, the undersigned, certify that all statements contained in this application are true to the best of my knowledge. I understand that any false answer or statements or implications made by me in this application or other required documents shall be considered sufficient cause for denial of employment or discharge. I agree to have a physical examination, including testing for substance abuse, conducted by a physician selected by XYZ and release XYZ from liability for nonemployment or termination of employment due to physical defects or history of substance abuse found in such examination. I hereby give XYZ the right to make a thorough investigation of my past employment, education, and activities and I release from liability all persons, companies, and corporations supplying such information. I indemnify XYZ against any liability which might result from making such investigation.

Additionally, I understand that nothing contained in this employment application or in the granting of an interview is intended to create an employment contract between XYZ and myself for either employment or for the providing of any benefit. If an employment relationship is established, I understand that I have the right to terminate my employment at any time and that XYZ has the right to terminate my employment at any time without advance notice or liability to me for wages or salary other than that earned by me prior to the termination of employment.

Signature ______________________ Date __________

Figure 7.2 *(continued)*

- Academic experience: quality of program; coursework in health, communications, business; grade-point average
- Personal experience: volunteering, traveling, other activities
- Leadership skills: officer of AWHP Student Club, fraternities/sororities, student government, athletics
- Writing skills: professional publications, school newspaper, project/thesis
- Speaking skills: professional presentations, Toastmasters International, debate team
- Consistency: continuous employment and/or education, extended periods of time in one position
- Appearance: attractive, well organized, concise, correct grammar and spelling

The Interview

After evaluating the applications for a given position, the manager invites the most promising applicants for interviews. This is an opportunity to screen each applicant further and to obtain more information that can be used to predict the applicant's success. For the applicant, the interview is a way to learn about the job and the organization and possibly to learn of other job opportunities in the organization.

In most organizations, the interview is conducted by the immediate supervisor of the prospective employee. The human resources department usually trains supervisors in the interviewing process. If the organization is small and human resources support is not available, supervisors can learn interviewing skills from their superiors.

A manager should go into an interview with a specific plan and well-defined objectives. This will allow the manager to maintain control of the interview and to obtain useful information. An effective interviewer will have a list of questions to ask and will be able to anticipate answers to maintain continuity throughout the interview. The interviewer should ask the same questions of each applicant for the position and then compare responses.

The interview should be conducted in a comfortable setting that is free from distractions. The manager should begin by putting the applicant at ease and attempting to develop rapport.

There are several ways to conduct an interview. It can be directive, nondirective, or a combination of the two. The directive approach is structured, with questions and their order predetermined and with little open discussion. The nondirective approach uses open-ended questions that give the applicant considerable latitude in responding. The interviewer then must respond and lead into the next question. Nondirective interviews encourage expression and allow the applicant greater involvement. However, this technique requires more planning and skill on the interviewer's part.

Whichever method is used, it is illegal to ask certain questions, such as marital status or family size. Managers who are conducting interviews should consult the human resources department for guidance.

Testing

In many organizations, the selection process includes various forms of testing. If used correctly, tests can provide valuable input for hiring decisions. However, it is essential that any test used be predictive of success in the specific job. Otherwise, the selection process may be challenged in court. Because the human resources department is responsible for testing programs, we will simply list some test types.

Among the common tests are

- intelligence tests,
- aptitude tests,
- proficiency tests,
- performance tests,
- interest inventories, and
- personality inventories.

In the health promotion area, some of these tests can be used to evaluate applicants. For instance, if a manager needs to hire aerobic instructors, applicants can be asked to complete a performance test by leading a group in aerobic exercise. Members of the selection team then can evaluate each applicant's test as a criterion for the hiring decision.

Reference Checks

The purpose of checking references is to verify information provided by the applicant and to obtain additional information from previous employers and other sources. References can be checked by telephone or letter. Reference checks can provide useful information about past job performance and reasons why the applicant is changing jobs. Note that it is illegal to check references and previous employment without the applicant's consent. Due to legal concerns, many employers refuse to give specific information beyond verifying that the applicant was employed by the organization between certain dates.

Sample Interview Questions

1. Scenario: You go into the CEO's office to explain your program. He/she blows smoke in your face and will not personally endorse the program. What do you say/do right then? What is your plan to gain his/her support in the future?
2. As a project manager, you are frequently called on to select vendors in the area of fitness (e.g., aerobic dance, etc.). What are the most important criteria in choosing the vendor?
3. How much liability insurance is adequate?
4. (If candidate doesn't know) Where would you find out?
5. How much training is adequate for aerobic dance exercise? (IDEA? ACSM? Degree?)
6. You have scheduled classes, with a minimum size of 15. Two weeks before the start date classes are not filled: Weight Control has 8 signed up, and Fitness has 12. What do you do?
7. At an interpretation session, a participant opens his/her HRP and finds another person's blood work inside. What do you do?
8. In looking at a group of employees to find potential action team leaders, how would you determine who would be your best candidate?
9. Do you believe in the use of incentives for a health promotion program? If yes, what do you think are the most effective incentives?
10. How many hours should it take to deliver a comprehensive health promotion program for 100 under one roof? Under four roofs?
11. What professional organizations do you belong to? (Note: Check résumé.) What would you join if you got this job?
12. If you could change your college preparation for this job, what would you have done differently?
13. What is a reasonable cost for a comprehensive health promotion program?

Preemployment Drug Testing

Because of the prevalence of illegal drug use and the high costs associated with employee drug use, an increasing number of employers administer preemployment drug tests. Although the cost of drug testing can be high, having a drug-free work force can save money in the long run.

The Job Offer

An organization's personnel policies will dictate how the hiring decision will be made and who will make the offer to the applicant. The hiring decision usually is made jointly by the human resources department and the manager of the department involved. In some organizations the human resources department forwards the names of recommended applicants for the manager's decision; in others the human resources department selects an applicant subject to the department manager's approval. The human resources department must approve the prospective employee's salary, benefits, and relocation allowance.

The actual job offer can be made by either the department manager or the human resources department. The offer should be made promptly to prevent the applicant from losing interest. The candidate should be contacted by phone and invited to join the organization, and a follow-up letter should be sent indicating the specifics of the job offer and setting a deadline for acceptance.

If the candidate rejects the offer, it may be useful to ask the reasons. The response may indicate the need to make changes in the selection process.

The candidate may make a counter offer. Before responding, the department manager should consult with the human resources department to determine which items of the offer are negotiable and to what extent.

Induction and Orientation

Most organizations have a procedure for induction and orientation of new employees. The procedure may include

- travel and relocation assistance and expense reimbursement;
- completion of tax, insurance, and other employment-related forms;
- explanation of the benefits package;
- explanation of the organization's policies and procedures;
- introductions to co-workers and supervisors; and
- orientation to job duties and responsibilities.

Responsibility for coordinating induction and orientation usually rests with the human resources department. The health promotion manager must orient new employees to the department and its operations. A skillfully conducted orientation provides valuable information to new employees that will improve their chances for success and ultimately will enhance the performance of the health promotion department.

In organizations where health promotion programs are an employee benefit, the human resources department should introduce new employees to health promotion operations in the induction and orientation process. This will enhance the department's visibility and increase participation in the programs.

STAFFING CHANGES

As organizations respond to technological, economic, and social change, their personnel needs also change. In the best cases these changes result from the predictable growth of a healthy organization. Often, however, they are a result of unforeseen circumstances that present a challenge to health promotion managers. For example, in addition to recruiting and selecting new employees, organizations may transfer, promote, demote, and terminate existing employees. Health promotion professionals must understand how these staffing changes are made so they can maximize the effectiveness of their programs.

Transfers

Transfers are movements of employees between jobs, units, shifts, or geographical locations. They can be initiated by the employer or the employee with approval of the employer. Some purposes of transfers are:

- filling vacancies in one area with employees from overstaffed areas,
- moving employees to positions more appropriate to their skills and interests,
- reassigning employees to areas where they can develop new skills, and
- moving employees to higher-priority positions as the result of redefinition of organizational goals.

Promotions

A promotion is an advancement to a position of greater responsibility, usually accompanied by increased pay, perquisites, and privileges and by greater potential for growth in the organization.

Demotions

A demotion is the opposite of a promotion. The employee is moved to a position with less responsibility, usually involving less pay, fewer perquisites and privileges, and less growth opportunity. Demotions result from inability to perform job functions or from staff reductions. Some demoted employees decide to resign.

Separations

When an employee and an organization sever their working relationship, a separation results. A separation can be a resignation, termination, layoff, or retirement. Whereas a layoff can be temporary, resignations, retirements, and terminations almost always are permanent. Most resignations and retirements are initiated by employees, and terminations are initiated by employers. Terminations result when employees violate laws or policies, or fail to perform at expected levels.

Succession Planning

Health promotion managers must always be prepared for staff changes by having competent replacements available for key positions. When possible, it is recommended that two qualified replacements be ready to meet unexpected needs in each key position. In health promotion this might mean having two employees who are trained to conduct exercise tests in the event the regular employee is unavailable, either temporarily or permanently.

Succession planning is a simple process in concept but may be complicated and lengthy in execution. The two basic steps in succession planning are identifying potential successors and preparing them for their roles, a process commonly known as grooming.

Identifying Successors

The first step in identifying successors is to consider the needs of the department and the organization in relation to the qualifications of the available candidates. Managers should select candidates not only for their ability to perform in the positions for which they are applying, but also for their potential for growth and development. Candidates who appear to lack some attributes needed for higher-level responsibilities may be considered for a vacant position but will probably not be an important part of a succession plan.

In the process of conducting regular performance appraisals, managers begin to form a clearer picture of each employee's capabilities. As skills and talents emerge, it becomes easier to match employees with the job descriptions for higher-level positions. As employees and managers jointly formulate employee development plans, individual interests also will emerge. Managers should consider all of this information in seeking to identify candidates for advancement and successors for superiors. Because higher-level positions may involve responsibilities that are outside the training and experience of health promotion professionals, managers must help promising employees prepare for their future roles.

Grooming

Employees may be groomed for higher-level positions in a number of ways. Formal training can be offered through tuition reimbursement plans for university study and through internal training programs and professional seminars and workshops. On the job, employees can be given assignments that expose them to new challenges and allow them to stretch their capabilities. Serving on committees, doing research for use by their superiors, and occasionally being delegated authority during vacations or leaves of absence can provide fertile ground in which an employee can grow.

Ideally, a manager should be able to sustain the loss of any employee with a minimal disruption in the flow of work and delivery of services. A promotion should never be delayed or denied because there are no viable candidates for replacement. Succession planning is the key to providing career growth opportunities for employees and to building an organization that can survive employee turnover.

Unpredictable Staffing Changes

In addition to predictable promotions and transfers, managers must deal with less predictable staffing changes, such as when employees become sick or injured, are terminated, resign, or take sudden leaves of absence. Although having a succession plan in place is a solution to the permanent loss of an employee, managers must develop contingency plans to cover short-term absences. These situations can often be handled through the use of temporary labor, interns, or overtime assignments for existing employees.

Temporary Labor

Temporary labor is one of the easiest solutions to a short-term employee absence. Unfortunately, health promotion professionals are not often found in the temporary labor market, except in areas where a college or university has a health promotion major or graduate-level program. In this case there may be an abundant source of entry-level health promotion personnel. Also, registered nurses and medical technicians may be used to perform health screenings and similar functions. Otherwise, temporary labor is most useful for replacing support staffers.

There are several factors to consider when deciding whether to use temporary labor.

- Does the assignment require specific training? It may not be practical to invest much training time in an individual who will be leaving in a few days.
- Does the assignment require specific skills that are outside of standard clerical job descriptions and subjective in nature? If the assignment requires meeting participants and learning their names quickly, a temporary employee may not be suited for the job.
- Can the temporary agency guarantee the same employee for the duration of the assignment? Retraining for even simple assignments may be inefficient.

Taking these into consideration, temporary labor is appropriate for such health promotion department activities as:

- Stuffing, collating, and packaging of materials
- Telephone coverage during absences and peak registration periods when the

intake of information is clearly systematized

- Non-technical functions such as height/weight measurement at a health screening, or staffing certain stations at a health fair
- Typical clerical duties, particularly when interpretation of data is not important, such as in transcribing taped presentations or dictation, or preparing mailing lists or labels

Interns

When an absence is expected to be of fairly long duration, the use of interns may be an excellent solution. This is especially true if the position is professional and the manager has some lead time to make arrangements. Interns generally outnumber internships and often are talented, hardworking people. Although requirements for internships vary among institutions, they tend to be flexible enough to allow a manager to use an intern in a legitimate professional role.

Interns do require supervision and expect to learn from the internship experience. Like entry-level employees, interns need strong direction and regular feedback. For this reason, the manager must weigh the advantages of using an intern to fill a temporary vacancy against the time that will be required to supervise the intern.

Overtime

Another possible solution for an unanticipated staff shortage is overtime for the remaining employees. Overtime offers a manager several advantages. First, training may be kept to a minimum because experienced employees are being used. Second, overtime can be instituted with very little notice. Third, when the short-term crisis is over, employees can easily return to their regular schedules.

There also are some disadvantages associated with overtime. First, it is probably one of the most expensive solutions if nonexempt employees are involved. The cost may be prohibitive for many health promotion budgets. If organizational policy permits, employees can be compensated for working overtime by being given time off with pay at a later date. Nonexempt employees probably will have to be paid overtime wages to comply with federal and state wage and hour laws.

A second disadvantage of overtime is that, for employees who are working overtime involuntarily, there may be a morale problem that can linger long after the overtime ends. Third, some state labor laws limit the number of hours part-time employees can work without being considered full time.

Managers must consider many factors when deciding whether to use temporary labor, interns, or overtime. It is advisable to anticipate the need and develop at least an informal plan. Managers should research vendors of temporary labor as well as sources of interns. They also should survey existing employees to ascertain their willingness and ability to work overtime when the need arises. Armed with this information, a manager can act quickly to cover an unpredictable staffing change.

EQUAL EMPLOYMENT OPPORTUNITY

When managing the staffing process, it is essential to be thoroughly informed about current legal issues and requirements. Of vital importance are the regulations and guidelines established and enforced by the federal Equal Employment Opportunity Commission (EEOC). In large organizations a human resources professional should be available to provide assistance with EEOC guidelines. Managers of small organizations may need to obtain assistance from employment agencies or government offices. The following three references are listed in the suggested resources and can be consulted for more information: Dogett & Doggett, 1990; Sedmak & Levin-Epstein, 1991; Shulman & Abernathy, 1990.

SUMMARY

Most of an organization's staffing functions are handled by the human resources department in cooperation with department managers. Because of the technical nature of health promotion, managers must have a thorough understanding of the staffing process so they can provide appropriate information to employees and obtain appropriate information from them. In small organizations where there is no human resources department, health promotion managers must consult with senior management or outside agencies.

Several factors are important in the staffing process:

- Personnel needs must be projected and plans made to fill them.
- A selection process must be established and followed.
- Alternatives for dealing with staffing changes must be considered.
- Equal Employment Opportunity Commission guidelines and other laws must be followed.

The effectiveness of a health promotion department depends not only on the selection and hiring of qualified employees but also on a manager's ability to appraise employees' performance and to provide opportunities for professional development. We consider these aspects of management in chapter 8.

KEY TERMS

grooming (p. 68)
separation (p. 68)
succession planning (p. 68)

SUGGESTED RESOURCES

Caruth, D. (1988). *Staffing the contemporary organization: A guide to planning, recruiting, and selecting for human resource professionals*. New York: Quorum Books.

Doggett, C., & Doggett, L. (1990). *The Equal Employment Opportunity Commission*. New York: Chelsea House.

Herriot, P. (1989). *Assessment and selection in organizations: Methods and practice for recruitment and appraisal*. New York: Wiley.

Lusterman, S. (1987). *The organization and staffing of corporate public affairs*. New York: Conference Board.

Sedmak, N., & Levin-Epstein, M. (1991). *Primer on equal employment opportunity*. Washington, DC: Bureau of National Affairs.

Shulman, S., & Abernathy, C. (1990). *The law of equal employment opportunity*. Boston: Warren, Gorham & Lamont.

Tyiner, A. (1990). *Organization staffing and work adjustment*. New York: Praeger.

Chapter 8

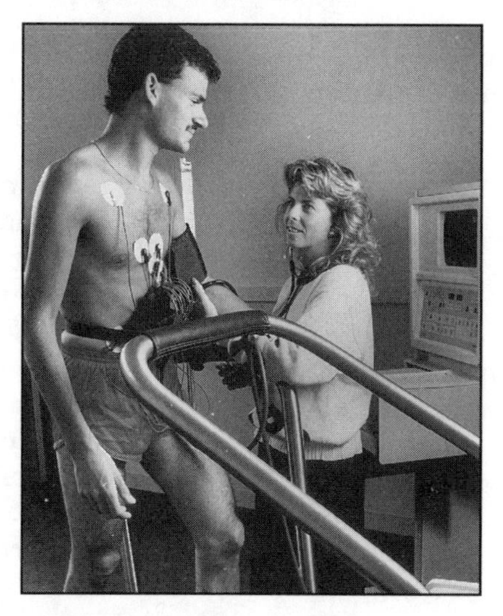

Appraisal, Training, and Development

To maintain a productive staff, supervisors must continually appraise employees' performance, train them on new equipment or programs, and help them develop their personal and professional skills. Training and development programs can prepare employees for new jobs, improve performance and productivity, enhance safety, develop teamwork, and provide retraining.

Appraisal is the evaluation of employee performance. In most organizations, informal appraisals are made on an ongoing basis. However, part of the process should be formal.

Training generally refers to the acquisition of knowledge or skills that have a specific application, such as how to use a new body-fat analyzer or how to teach a smoking cessation class. In contrast, development refers to acquiring knowledge or skills that have a broader application, such as operational management, management of corporate culture, organization theory, or supervision.

PERFORMANCE APPRAISALS

A formal appraisal process should include a periodic, systematic evaluation of all employees at predetermined intervals throughout their tenure. In this process managers will obtain important information for making decisions on pay increases, promotions, transfers, selection, correction, training, and development. It is appropriate to conduct appraisals

- before hiring for selection purposes,
- during the first several days of employment for correction purposes,
- every six months for performance review,
- every year for salary review, and
- every three years for promotion evaluation.

As noted earlier, informal appraisal is an ongoing process, whether at the conscious or subconscious level. The issue is not whether to appraise, but how and how often and how to make appraisals a constructive process.

In most large organizations, performance appraisal procedures and forms are developed by the human resources department. In smaller organizations, managers must develop these procedures and forms with guidance from superiors or consultants.

The appraisal process can involve two kinds of evaluation: merit rating and appraisal interviews.

Merit Rating

Merit rating is a systematic process by which supervisors rate the performance of their employees. Several methods are used that are based on criteria such as finishes work in a timely manner, works well with others, and takes individual initiatives.

- The graphic rating-scale method uses a list of traits of good employees. For each trait, the rater checks the appropriate descriptor in ranges from superior to inadequate, excellent to poor, and similar scales.
- In the rank-order method, the supervisor ranks all employees from best to worst.
- The forced-distribution method uses scales similar to those in the rating-scale method, but the supervisor must assign fixed percentages of employees to the middle and both ends of each scale.
- In the paired-comparisons method, the supervisor compares each employee with every other employee on each item.

Each merit rating method has distinct advantages and disadvantages; these are summarized in Table 8.1. The appropriate method for any situation will depend on the needs of the organization, department, and manager, and the resources available to complete the process.

The major problem with merit rating is that the reliability of ratings depends on the rating skill of the supervisor (rater) and the rating instrument. Reliability improves when raters are well trained. Reliability also improves when the instrument uses descriptive phrases on the rating scales, for example, "goes out of the way to cooperate" rather than "above average." Finally, reliability improves with forced choices, where the rater must indicate a positive or negative response and no "indifferent" response is available. This is particularly true if there are a few as opposed to many items on the instrument.

Despite the problem of reliability, merit rating can be an effective tool if an instrument that considers traits necessary to perform the job is used and raters are well trained. In large organizations, rating systems are selected and implemented by the human resources department. Managers must learn how to use the system correctly.

Appraisal Interviews

The purpose of conducting an appraisal interview is to provide feedback on the results of the employee's merit rating. The supervisor's objectives may include

- a general assessment of behavior and performance,

Table 8.1
Advantages and Disadvantages of Merit Rating Methods

Method	Advantages	Disadvantages
Graphic rating scale	Easy to use	Different raters rate differently
Rank order	Good for differentiating among employees	Hard to use with large numbers of employees
Forced distribution	Good for differentiating among employees	Different raters rate differently; hard to use with large numbers of employees
Paired comparisons	Best for differentiating among employees	Hard to use with large numbers of employees

- encouragement to continue present behavior and performance,
- a statement of the need to change behavior or performance, or
- a formal warning that behavior and/or performance are unacceptable.

Before the interview, the supervisor should direct the employee to perform a self-appraisal. The responses can be used as a basis for discussing problems and possible solutions.

In the appraisal interview, the supervisor should begin by reviewing the employee's performance, then discuss the employee's self-appraisal. The interview next should focus on ways to improve performance and on the establishment of goals and objectives. The supervisor should document the interview in a letter to the employee that summarizes the discussion.

STAFF TRAINING

To build and maintain a staff of competent employees, managers must develop an effective training program. This is particularly important in the health promotion field, where there are frequent innovations in equipment, programs, and techniques. Additionally, training is needed to develop the skills to meet department and organizational goals.

Determining Training Needs

A training program should be offered only when a specific need exists. Specific problems must be identified and analyzed to determine whether training will solve the problem, and then appropriate programs must be developed or found. There are several methods of determining training needs (see Table 8.2).

Training Methods

There are many training methods. A partial list and the situations for which each is best appears in Table 8.3. The approach that is selected should meet the training needs that have been identified and should be a method with which the manager is comfortable.

Authority and Accountability

The overall responsibility for training usually belongs to the human resources department. In larger organizations there may be

Table 8.2
Methods of Determining Training Needs

Method	Function
Organizational analysis	Determines which training efforts should be emphasized throughout the organization
Organizational climate analysis	Determines employees' attitudes, feelings, beliefs, and opinions
Operations analysis	Determines which training efforts should be emphasized for individual employees based on specific jobs
Personnel analysis	Determines behavioral traits employees need to perform jobs

Table 8.3
Methods of Training Employees

Method	Appropriate situations for use
Lecture	Explain a large amount of information to a large number of people
Coaching	Explain information to one person or a small group where immediate feedback and reinforcement are beneficial
Group discussions	Stimulate analytical thinking with little new information
Case studies	Stimulate problem solving using realistic situations
Projects	Stimulate independent work and creativity
Sensitivity training	Develop awareness of needs and emotions of others
Transactional analysis	Provide insight and perspective for effective response to behavior of others

a separate training department or a training unit within the human resources department. However, department managers are in the best position to determine specific training needs and to provide skill training. Managers must be familiar with the methods for determining training needs and know how to train employees in the specific skills needed in health promotion at a specific worksite, such as using computer software and specific exercise equipment.

EMPLOYEE DEVELOPMENT

As noted earlier, training programs teach the skills required to perform specific jobs. Development programs focus on enhancing management skills needed to achieve broader organizational objectives. For example, managers of operating units might attend a session on how to enhance employee communications or establish a management by objectives program. Development programs also may be used to groom individuals for positions of greater responsibility.

There are many types of development programs. Some are essentially the same as training programs except that they have different objectives. Those commonly used in health promotion are listed in Table 8.4.

AUTHORITY AND ACCOUNTABILITY

In most organizations, the responsibility for employee development belongs primarily to the human resources department. With organizational objectives in mind, these specialists determine development needs, recommend and develop programs, and follow up. Immediate supervisors may decide which of their employees will attend what sessions.

For health promotion managers, the human resources department can be a resource for identifying appropriate employee development programs or providing ideas and assistance for creating development programs within the department. Another resource is professional associations. Organizations that sponsor annual conferences in the field of health promotion include

- American Alliance for Health, Physical Education, Recreation and Dance (AAHPERD),
- American Council on Exercise (ACE),
- *American Journal of Health Promotion*,
- Association for Worksite Health Promotion (AWHP),
- Canadian Association of Health, Physical Education and Recreation,
- International Association of Fitness Professionals (IDEA),
- National Wellness Institute,
- Wellness Councils of America (WELCOA), and
- Wellness in the Workplace.

Table 8.4
Methods of Employee Development

Method	Function
Coaching	Explain information to one person or a small group where immediate feedback and reinforcement are beneficial
Counseling	Give advice to an employee (one component of coaching)
Delegation	Stimulate independent work and creativity by assigning managerial tasks and providing limited supervision
Continuing education	Provide knowledge and skill development outside organization; e.g., conferences, seminars, university courses

SUMMARY

Appraisal, training, and development activities should be well supported by the human resources department. However, health promotion professionals must understand these processes to maximize the productivity of employees.

Appraisal, training, and development are essential to building and maintaining a strong health promotion staff. Equally important to a program's success is the ability to communicate with other department managers, senior management, participants, and the public. In chapter 9 we address all these aspects of communication.

KEY TERMS

appraisal interview (p. 74)
merit rating (p. 74)
performance appraisal (p. 73)

SUGGESTED RESOURCES

Bard, R. (1987). *The trainer's professional development handbook*. San Francisco: Jossey-Bass.

Craig, R. (1987). *Training and development handbook: A guide to human resource development*. New York: McGraw-Hill.

Goldstein, I. (1986). *Training in organizations: Needs assessment, development, and evaluation*. Monterey, CA: Brooks/Cole.

Lynton, R. (1990). *Training for development*. West Hartford, CT: Kumarian Press.

Mitchell, G. (1987). *The trainer's handbook: The AMA guide to effective training*. New York: AMACOM.

Nadler, L. (1989). *Developing human resources*. San Francisco: Jossey-Bass.

Robinson, D. (1989). *Training for impact: How to link training to business needs and measure the results*. San Francisco: Jossey-Bass.

Smith, B. (1987). *How to be an effective trainer: Skills for managers and new trainers*. New York: Wiley.

Spaid, O. (1986). *The consummate trainer: A practitioner's perspective*. Englewood Cliffs, NJ: Prentice-Hall.

Chapter 9

Communication Management

One of the fundamental components of management is effective communication. Because management involves directing the work of others, the ability of managers to communicate is paramount to their success. Communication is a field of its own, and entire books have been written on individual aspects of the subject. This chapter discusses some of the basic concepts and their relation to health promotion, such as types of communication, types of meetings, and delivering presentations.

FUNDAMENTALS OF EFFECTIVE COMMUNICATION

Communication in management has two basic functions: to link components of an organization by transmitting and receiving information (after which the functions of management can be executed) and to establish and promote human relations by building good interpersonal relationships. Together these functions should result in positive employee actions.

Whatever the purpose of communication, it includes four fundamental components:

- Source (the speaker, writer, or artist)
- Message (what the source is trying to transmit)
- Symbol (used to transmit the information; can be in the form of words, signs, or gestures)
- Receiver (the listener, reader, or viewer)

Every component is important to management communication. To be effective, management communication must be a two-way exchange of information. After the message is sent, the source must evaluate the response of the receiver to determine if the message was understood.

The two major types of organizational communication are formal and informal. The patterns of formal communication can be upward, downward, or lateral; informal communication can occur by channels like the grapevine and cliques.

FORMAL COMMUNICATION PATTERNS

The organization's structure (which was discussed in chapter 4) outlines the formal communication channels. In a review of Figures 4.1 through 4.3 you can see that information can flow formally in two directions, vertically and horizontally. This results in communication that moves down, up, or laterally.

Downward Communication

In the traditional model downward communication flows from top management to the bottom of the organization, informing subordinates of plans, policies, and directives. If information is clear, more efficient operation is the result. Informed employees tend to be more trusting of management and more willing to support the organization's goals. They also tend to feel more a part of the organization, which can improve their performance.

In health promotion, effective downward communication flows from the health promotion professional to program participants.

There are various methods for communicating information downward, including these:

- Direct personal contact
- Orientation materials and programs
- Bulletin boards and posters
- Written materials (policies, procedures, reports, handbooks)
- Newsletters
- Paycheck stuffers
- Letters
- Audiovisuals (videos, slides, public address)

To maximize downward communication, whether from senior management to employees or from the health promotion manager to program participants, there are several basic guidelines.

First, it is the sender's responsibility to make sure the message is clear. A message is clear only if the sender and the receiver speak the same language or interpret symbols the same way. Because an individual's interpretation of symbols is subjective and is a function of education, experience, and background, the success of downward communication is strongly dependent on the manager's ability to assess how symbols will be interpreted by the receivers.

Second, the manager must determine the mode of communication, such as written, oral, or visual presentation. Information usually is communicated most effectively when two or more symbols are used. For example, an oral presentation with diagrams is easier for many people to understand than is a written presentation or a lecture without visual aids.

A demonstration, which uses oral and visual symbols, can be an effective way to communicate information to health promotion program participants who must learn to use exercise equipment.

When transmitting information orally, the sender should do the following:

- Use concrete and specific words.
- Use words the receiver is likely to understand.
- Define words that may not be understood.

Third, the way in which the message is presented strongly affects the success of downward communication. Here are some guidelines:

- Use repetition if information has not been completely transmitted.
- High-ranking managers should make presentations (the higher the rank, the greater the effect).
- Managers should make important presentations orally rather than in writing.

Fourth, timing is critical. Managers should be certain that

- information is disseminated in a timely fashion to prevent rumors,
- supervisors always obtain information before their subordinates,
- all employees of equal rank are informed at the same time, and
- the information does not conflict with any information that was previously communicated.

Upward Communication

For there to be a two-way flow of information, there must be communication from the lower levels of the organization toward senior management. This is called upward communication. Although many authoritarian managers prefer one-way, downward communication, it is through upward communication that management receives feedback from employees and health promotion managers receive information from program participants.

Feedback is valuable because it indicates how downward communications are being accepted, gives employees or participants a sense of belonging, allows them to contribute ideas for improving operations, and helps managers anticipate problems before they develop.

Upward communication can take many forms:

- Surveys, such as needs assessments
- Employee committees, including advisory groups
- Suggestion systems
- Discussion meetings between management and employees
- Formal grievance procedures
- Union representation and publications

All of these channels except the last are formal lines of communication established by management to encourage employee feedback. In health promotion, an important form of upward communication is the advisory group of constituents who provide feedback to the health promotion professionals.

Upward communication should be a systematic process in which information moves through the levels of management to the person responsible for making the decision. For this process to work, managers must be good listeners. Listening well includes these elements:

- Being receptive to criticism
- Being attentive when communicating with employees

- Paying attention to all employees, not just those with whom the manager agrees
- Being willing to acknowledge mistakes and make changes

After listening effectively, the manager must follow up with an adequate response. This could be

- communicating employee concerns to higher levels of management,
- taking action, or
- explaining why nothing can be done.

Managers should never imply or promise they will take some action and not follow through. This inevitably leads to resentment on the part of employees.

Lateral Communication

When information moves horizontally through the organizational structure, there is communication between management levels of equal rank. This promotes teamwork through coordination of activities among departments. This is an important communication channel for health promotion managers because other department managers can provide valuable assistance and advice (pages 30-33).

Lateral communication also involves the flow of information to and from people outside the organization. This can help build strong community relations and provide access to information not available within the organization, such as competitor's product or service characteristics identified by networking with people outside the organization. It is beneficial to make the community aware of the health promotion services the employer offers to employees. Also, community resources, such as hospitals, health agencies, and government health departments, can provide valuable health services and information to share with employees.

There are several ways to enhance horizontal communication both within an organization and with outside entities. Some of these are the following:

- Conferences, meetings, and seminars
- Training sessions
- Interdepartmental coordination of functions
- Meetings with groups outside the organization
- The collective bargaining process

INFORMAL COMMUNICATION CHANNELS

Communications that take place independent of the organizational structure are informal. Whereas formal communications often are written, informal communications are almost always oral. This is probably because people do not want a permanent, written record of these communications. Also, informal communications generally do not follow any consistent patterns.

There are many kinds of informal communication. We will discuss two: the grapevine and cliques.

The Grapevine

The informal communication that winds through the formal structure like a grapevine (hence its name) exists in all organizations. The volume of information that flows through the grapevine is usually related to the timeliness and quantity of formal information. When senior management fails to keep employees informed formally, more information is communicated through the grapevine.

Typically, the information has some basis in fact but becomes misconstrued as it is retold. This results in gossip, rumors, and conjecture that can harm the organization.

Managers can minimize the grapevine's potential for harm by maintaining effective, formal, two-way communication.

Cliques

Like the grapevine, the informal formation of groups of employees with similar interests exists in all organizations. These groups are typically called cliques. Although cliques can actually be a positive force, it is important to make sure that a clique does not become more important to its members than the organization. Again, effective, formal, two-way communication can minimize this problem.

Because a clique usually forms around one or more natural leaders, cliques can be useful in promoting health promotion activities. If the leaders of a clique believe in and participate in health promotion efforts, the members of the clique are more likely to participate as well.

Now that we have discussed various kinds of formal and informal communication channels, we will examine two specific types of formal communication in which health promotion managers might be involved: conducting meetings and giving presentations.

CONDUCTING MEETINGS

Around the world, business people are spending more and more time in meetings. The purpose of most meetings is to solve problems. Typical participants are executives, managers, and people with relevant technical expertise. To make the best use of these people's valuable time, managers must be able to conduct effective meetings.

Types of Meetings

Meetings can be categorized in many different ways. We will classify meetings by how they are conducted and how they are scheduled. Meetings can be formally or informally conducted, and they can be scheduled on a regular basis or called for special purposes.

Roundtable and Agenda-Driven Meetings

A meeting can be informally or formally guided. The determining factor is a well-defined agenda. Agenda-driven meetings have an agenda that indicates what will be discussed and in what order (see Figure 9.1). A roundtable meeting is less rigid and allows for open discussion. The chairperson is responsible for facilitating the discussion and keeping order.

Regularly Scheduled and Special-Purpose Meetings

Some meetings take place on a regular basis, and some meetings are scheduled when a specific need arises. Department and advisory council meetings typically are regularly scheduled. In the health promotion department, a meeting may be held once every two weeks and an advisory council meeting once a month. Special meetings may be called to address these situations:

- Learning to use a new piece of equipment
- Scheduling during a possible strike
- Hiring a new employee
- Planning a new facility
- Evaluating a recent drop in program participation

Preparing for a Meeting

The more carefully a meeting is planned, the more productive it will be. To plan a successful meeting, follow these steps:

- State clearly the purpose of the meeting.

MONTHLY STAFF MEETING

The monthly meeting for the entire health promotion staff is scheduled as follows:

Tuesday, April 2
9 - 10:30 am
Headquarters building
Conference room 7N01

As always, be prepared to present a brief activity summary for your facility. Bring handouts and have overheads prepared. You will be limited to a 5-minute overview. The meeting agenda is as follows:

- General announcements
- Facility activity reports
- National Heart Month update
- Fall health screening
- Spring noontime outside activities restrictions
- Second-quarter committee assignments
- Committee organizational meeting
- Low birthweight baby project
- New hire orientation issues
- Open forum

This is a fairly ambitious agenda for the time we have. However, if you have any other important agenda items, let me know on Monday and we'll try to work them in.

P.S. *It's Mod A's turn for treats.*

Figure 9.1 Establishing a meeting agenda ensures that all topics are covered and keeps the meeting moving.

- State specific, measurable objectives that will achieve the purpose.
- Identify the decision maker(s), i.e., the group or the leader.
- Determine the meeting format, i.e., roundtable or agenda-driven.
- If appropriate, develop an agenda.
- Choose the participants.
- Notify participants of their responsibilities.
- Choose the chair and the recorder.
- Establish the time, place, and duration of the meeting.
- Identify audiovisual and refreshment needs.
- Create a plan for following up on action items (specific tasks identified at the meeting).
- Develop a method for evaluating the meeting.
- For a large, formal meeting the leader's opening remarks should be prepared

and practiced. This speech sets the tone for the entire meeting.

If these steps are completed, the meeting should be successful and productive, with participants accomplishing their objectives in a reasonable amount of time.

Running the Meeting

A meeting not only must be well planned; it also must be well run. Responsibility for running the meeting lies with the leader. If the organization has established procedures for conducting meetings, the leader should follow them. A widely used procedure for formal meetings of boards and associations is found in *Robert's Rules of Order* (Robert, 1989). These principles can be adapted for business meetings.

The Leader's Role

The leader has overall responsibility for the effectiveness of a meeting. An effective leader should do the following:

1. Have a plan (agenda) and follow it
2. Begin the meeting with a short orientation speech to:
   - State the objectives and procedures for the meeting
   - Provide factual information to be used as a basis for discussion
   - Clarify what topics are and are not open for discussion
   - Review the agenda
   - Select person to record the minutes
3. Maintain an environment where all members feel comfortable making contributions to the group
4. Maintain clarity by
   - Providing or obtaining clear, accurate summaries
   - Providing or obtaining clarification of ambiguous statements
   - Encouraging review of all generalizations that are not supported by facts
5. Provide the opportunity for everyone's opinion to be considered
6. Minimize conflict that is not directly related to the problems or tasks at hand
7. Provide input only when absolutely necessary and spend time monitoring and controlling the meeting
8. Begin and end the meeting on time
9. Conclude the meeting with a review of actions and assignments

The Participants' Role

Participants as well as the leader have responsibilities for making a meeting productive. All participants should do the following:

- Organize their thoughts before speaking, possibly in writing
- Speak forcefully
- Speak only when they have relevant comments
- Make only one point at a time for clarity
- Support their contributions with solid evidence
- Use positive nonverbal communication by looking the speaker in the eye and not using negative facial expressions
- Listen attentively to all comments

Adherence to these guidelines will promote the active participation that results in a successful meeting.

Evaluating the Meeting

The information obtained by conducting post-meeting evaluations can be useful in improving future meetings. The evaluation process need not be time consuming. Some simple approaches follow.

Meeting participants can be asked to complete a short evaluation form. The form may request an evaluation of the leader, the participants, and the general effectiveness of the meeting. Space should be provided for suggestions on how future meetings could be improved. The leader should review information, summarize it, and distribute a report to the participants.

Another way to evaluate a meeting is to invite an observer who is knowledgeable about meeting dynamics. The observer should take notes and make suggestions for possible improvements.

PREPARING AND DELIVERING PRESENTATIONS

Presentations are an important form of communication in health and business, and managers usually are evaluated on the quality of their presentations. In health promotion, the program itself is evaluated on the basis of the manager's presentations. Health promotion managers use presentations to explain ideas and convince others of the ideas' validity and worth. Presentations are an important tool in both formal and informal communication. Managers who want to improve their public speaking skills may benefit from attending training sessions, such as Toastmasters International.

To deliver an effective presentation, a manager must follow four steps:

- Determine the objective of the presentation.
- Evaluate the audience.
- Develop the presentation.
- Prepare the delivery.

The Objective

A worthwhile presentation has a specific purpose. The purpose should be clearly stated in the form of an objective. The objective should not be to deliver a great presentation, but to induce the audience to respond in a specific way. That is, the objective should state how the presenter wants the audience to behave as a result of the presentation.

Once the objective has been determined, it should be the basis for the preparation and delivery of the presentation. The objective can be used to

- narrow the scope of the presentation;
- identify the theme of the presentation; and
- decide on the mode of delivery, i.e., formal versus informal.

The Audience

A key factor in the success of a presentation is the presenter's understanding of the group for which the message is intended. To effect the intended behavior change in the audience, the presenter should know what level of interest the audience has in the topic, how knowledgeable the audience is about it, and what attitudes they have toward it. By evaluating the audience in advance, the presenter can determine the appropriate levels of detail and persuasion.

There are several ways to obtain information about an audience. First, the presenter can review of list of attendees and their titles and departments. The presenter will know some of the people on the list and will have some idea of their knowledge level and whether or not they are likely to support the presentation's objective. For people the presenter does not know, information about their knowledge level and attitudes can be obtained by asking them directly, by requesting all attendees to complete a questionnaire, or by asking colleagues who do know these people.

The Presentation

Once the presenter has completed an assessment of the audience, the next step is to develop the presentation. Here are some guidelines:

- Communicate all relevant information (it is better to say too much than not enough).
- Make the level of detail understandable to the least knowledgeable participant.
- Use supporting evidence as appropriate (again, too much is better than not enough).
- Develop two presentations if there are two extremes of knowledge in the audience.
- Use examples to change attitudes (sometimes several examples are needed).
- Use powerful arguments and plan them carefully.
- Do not use personal pronouns (me, myself, I, mine): The focus should be on the impact of ideas rather than the source.
- Use testimony of other influential people.
- Attribute ideas to others to help change attitudes.

Because many presentations in the health promotion area revolve around lifestyle changes, both knowledge and attitudes are extremely important. Following the preceding guidelines will help a presenter effectively convey the right message to the audience.

The Delivery

After identifying the objective, evaluating the audience, and developing the presentation, the presenter can prepare the delivery. The mode of delivery chosen should be consistent with the objective and the audience. Several modes of delivery are summarized in Table 9.1.

Table 9.1
Ways to Deliver Presentations

Type	Function
Review	Evaluate progress and performance
Proposal	Present new ideas or modifications of old ideas
Informational	Present explanations of job functions
Motivational	Attempt to change attitudes

When preparing the delivery, it is important to keep the objective and audience clearly in mind. A presentation can be divided into three parts: introduction, body, and conclusion.

Use the first few minutes to set the stage for the body of the presentation. This is the time to present an overview to the audience and establish rapport with attendees. In the introduction, do the following:

- Briefly present the background of the speaker to establish credibility.
- Briefly state the purpose of the presentation.
- Briefly outline the extent of the presentation (because no presentation can cover every aspect of a topic).
- State any expectations of the audience that will help listeners benefit from the material to be presented.
- Explain how the presentation will benefit the audience.

If the objective of the presentation is to induce the audience to support certain recommendations, spell out the criteria to be used for making the decision. This will give the audience a basis for evaluating the information to be presented.

The body of the presentation is the actual information to be conveyed. This section must be well organized. Focusing on the theme of the presentation, the presenter should develop an outline of the major topics and then expand on it in detail.

In preparing a conclusion, summarize the main points of the presentation and suggest ways in which the audience can act to meet the stated objective. Where appropriate, specify the desired action, for example, signing up for a stress management course, changing cafeteria food selections, establishing an advisory committee, or purchasing new exercise equipment.

SUMMARY

The value of communication management should not be underestimated. Health promotion professionals must develop their skills in this area to create presentations that achieve their objectives. To communicate effectively, health promotion managers should do the following:

- Study the organizational chart to identify formal communication channels.
- Consider all informal communication patterns.
- Plan organized, productive meetings.
- Plan effective, organized presentations and attend training sessions to improve public speaking skills.

Communication plays a role in every major management function, from hiring and training employees to marketing, budgeting, and planning. Another function that involves every aspect of management is computer operations, which is the subject of chapter 10.

KEY TERMS

cliques (p. 83)
grapevine (p. 82)

SUGGESTED RESOURCES

Belanger, S. (1989). *Better said and clearly written: An annotated guide to business communication sources, skills, and samples*. New York: Greenwood Press.

MacStravic, R. (1986). *Managing health care marketing communications*. Rockville, MD: Aspen Systems.

Meuse, L. (1988). *Succeeding at business and technical presentations*. New York: Wiley.

Mosvick, R., & Nelson, R. (1987). *We've got to start meeting like this! A guide to successful business meeting management*. Glenview, IL: Scott, Foresman.

Penrose, J. (1989). *Advanced business communication*. Boston: PWS-KENT.

Perret, G. (1989). *Using humor for effective business speaking*. New York: Sterling.

Robert, H. (1989). *Robert's rules of order*. Nashville: T. Nelson.

Tebeaux, E. (1990). *Design of business communications: The process and the product*. New York: Macmillan.

Chapter 10

Computer Applications

Computers are an integral part of virtually all business operations, including health promotion programs. For this reason, health promotion professionals must have a general understanding of modern computer technology and know how to apply it effectively to health promotion operations. This requires knowledge of available hardware and operating systems, as well as software for a variety of standard and specialized applications.

In this chapter we will describe the two main types of computer operating systems and outline the features of standard business software programs as well as applications specifically designed for use in health promotion.

HARDWARE AND OPERATING SYSTEMS

In today's business environment, there are two kinds of operating systems: MS-DOS (IBM based) and Macintosh.

MS-DOS Systems

MS-DOS (Microsoft Disk Operating System) is the predominant operating system in business today. It is used to run IBM personal computers or so-called IBM compatibles or "clones." The hardware is available in a variety of configurations with different levels of memory, disk capacity, microprocessors, and monitor types. Options include a hand-controlled "mouse," a modem, an external drive, and other equipment.

A wide variety of software can be used with MS-DOS; both standard and specialized applications will be described later in the chapter. With its printed menu screens and keyboard controls, MS-DOS seems to be best suited for "left-brained" individuals who work well with written language and logic. However, in response to competition from Apple Computer, the Windows type of software that uses "dropdown" menus, icons, and a mouse is becoming the standard for IBM computers and clones.

Macintosh Systems

The other operating system is that designed for Macintosh hardware made by the Apple Computer Company. Like IBM machines, the "Mac" is available in many configurations and with a variety of optional equipment. While not the predominant system for business applications, Macintosh is gaining in popularity because of its ease of use.

The Macintosh system, which uses a mouse and icons instead of printed menus, is generally easier for "right-brained," visually oriented individuals to learn. Although most standard software applications are available for the Macintosh, the system is best known for its graphics and desktop publishing applications.

By the time they enter the job market, most health promotion professionals will have been exposed to both types of systems and will have developed a preference. In addition to personal preference, the selection of a system will depend on the specific application and the availability of suitable software. In some organizations, managers may not have a choice of systems and will have to use the equipment specified by senior management.

STANDARD SOFTWARE APPLICATIONS

As computers have evolved and become integrated into the business environment, a number of software applications have become standard. Among these are word processing, spreadsheets, database management systems, desktop publishing, and graphics programs. We will describe each of these in the sections that follow.

Word Processing

Word processing programs have replaced the typewriter for almost all typing applications except filling out forms. In a health promotion program, word processing is used for preparation of documents, mail merge, newsletters, flyers, and posters.

Documents

Document preparation is the most basic function of word processing. It allows the user to create documents, modify them, and save them on a disk. Many programs also have spell checking and thesaurus functions. With newer programs, it is possible to create documents that can be converted into formats for faxing by modem to remote locations.

It is useful to create a library of standard documents or forms that are used repeatedly in the administration of a health promotion program. Examples are activity registration forms, disclaimers and waivers, activity confirmation letters, vendor contracts or statements of work, work plans, and articles to be used in newsletters. Much time and labor can be saved by using these as "boilerplate" documents that can be modified to suit a specific circumstance. Many of these documents can be used with a mail merge function.

Mail Merge

Mail merge allows the user to merge a form

document with a database of names, addresses, and other information to create the appearance of personal letters. For example, a letter confirming health screening appointments can be created with codes inserted for the name, mail station, and time and date of appointment. The document is merged with a database of recipients' names, mail stations, and appointment times and dates. The computer then prints the text for each letter inserting personal data when prompted by the codes.

Mail merge also can be used to prepare mailing labels or envelopes, registration forms, telephone lists, and rosters. Mail merge must be used with a database. Some word processing programs have a built-in database function. Others can take data from a separate database and combine it with text created in the word processing program.

Newsletters

Although desktop publishing programs offer more flexibility, word processing programs can be used to create attractive newsletters. Most programs offer a choice of type styles and sizes so the user can create headlines and captions. A disadvantage is that word processing programs do not easily allow for the incorporation of graphics directly into the newsletter. For more sophisticated publications, a desktop publishing program is necessary.

Flyers and Posters

Word processing programs also can be used to create flyers and posters. For multisite programs that use compatible software, generic flyers and posters can be created centrally and personalized at each individual site. Again, if a more sophisticated product is needed, the appropriate choice is desktop publishing.

Spreadsheets

Spreadsheet programs consist of a grid of "cells" identified by column and row numbers or letters. Each cell can contain text, numerical data, or formulas. Individual cells can be formatted to display data in any of a number of formats. Because the basic form of a spreadsheet is a grid, it is well suited for any data that is to be displayed in columns and rows, such as a table of numbers or a roster of names and phone numbers.

The most valuable feature of spreadsheet programs is their ability to make many complex calculations in an instant. This makes them particularly suitable for financial applications and a variety of recordkeeping and analysis functions. Most spreadsheet programs also have some database and graphics functions.

Financial Applications

Financial applications are the function most often associated with spreadsheets. They are used for forecasting, building, and tracking budgets. They are also useful for tracking receipts from participant fees and inventories. Like word processing programs, spreadsheet programs allow the user to create "boilerplate" budgets or other forms with predetermined formulas. As data are entered, the program performs the desired calculations automatically. Many programs have an external import feature that allows each facility in a multisite program to create an individual budget and permits the manager to consolidate them into an overall budget without entering all the individual data.

Data Processing

Any kind of numerical data can be recorded and analyzed on a spreadsheet. Examples of applications are activity attendance, enrollment, facility use, salary forecasting, and similar tasks. If data can be recorded in columns and rows as would be done on graph paper, it probably can be done more quickly and easily on a spreadsheet. In fact, many users design spreadsheets on graph paper.

In addition to basic mathematical calculations, modern spreadsheets can do complex tasks such as calculating mortgage rates and making statistical computations. This capability enables the manager of a relatively small health promotion program to conduct research or apply sophisticated analysis to program data for reporting purposes.

Database Applications

Spreadsheets have some applications as databases. If one considers each column a field and each row a record, it is possible to sort and select data on the basis of predetermined criteria. For example, a health promotion manager might maintain a list of participants as records with their program interests as fields. Using the database capability, the manager can direct the spreadsheet to select all participants who indicated nutrition as their primary program interest and do a follow-up promotion specifically targeted to those people.

Spreadsheets have limitations in size and speed when used as databases. If the amount of data causes the program to run slowly, a true database application is recommended. Databases will be discussed in more detail later in the chapter.

Graphics Applications

Full-featured spreadsheet programs can convert tables of numbers into graphs and charts. This feature allows a manager to use one spreadsheet both to track data and to create charts to report the data to management. When the data are changed, the charts are automatically updated as well.

The graphics feature of a spreadsheet program is adequate for many needs. However, there are limitations with respect to format choice and printer selection. For more sophisticated applications produced on laser printers with color features, for example, it may be necessary to use a desktop publishing or graphics program.

Database Management Systems

True database management systems, often called relational databases, allow the user to collect data and retrieve them, manipulate them, and create reports on the results. As explained in the section on spreadsheets, there are two basic elements in a database: records and fields. A record is one entity, such as a participant, to which any number of attributes can be recorded. These attributes, such as demographics or responses to a questionnaire, are called fields. Once a record has been created in a specific database, any number of fields can be added. Fields added to one record must be added to all records, although it is not necessary to put any data in a field.

In a health promotion program, database management systems are most useful for tracking participant activity and targeting segments of the participant population.

Participant Tracking

In this basic function, the records consist of all participants or possibly eligible participants. Possible fields include

- name,
- age,
- sex,
- job type,
- shift,
- phone number,
- mail station,
- whether employee has a health risk assessment on record,

- primary risk area,
- rankings of responses to an interest survey, and
- whether employee completed or dropped out of an activity.

Once the database has been set up and the data entered, the manager can extract data for program planning. Using some of the fields listed above, the manager could request a list of all participants and their mail stations who meet criteria of

- over 40 years old,
- male,
- working first shift,
- have indicated interest in smoking cessation,
- have never attended a program-sponsored smoking cessation class, and
- attended one Great American Smokeout activity.

The list of names and mail stations could be formatted into a distribution list or mailing labels. The manager could then target this group with a specific promotion.

Targeting Populations

As will be explained in Part III on marketing, a basic marketing technique in health promotion is segmentation. For a population of more than 1,000, a database system is necessary for conducting targeted programming to specific segments of the population. An easy way to build a health promotion database is to transfer from an existing human resources disk the names and basic demographics of the entire eligible population.

Individual fields can be updated as activity progresses. For example, if 20 individuals register and attend an activity, that data can be entered at the conclusion of the activity or with the records from other activities. The objective should be to maintain an accurate, current database that can be used for target marketing. More information on targeting populations is presented in Part III.

Desktop Publishing

Desktop publishing is a specialized application used to create newsletters, posters, and flyers. Much more flexible and versatile than word processing, desktop publishing programs offer a wide variety of type styles and sizes, electronic clip art, scanned images, and other features. Most programs are used with a mouse and can import text from word processing files.

The most sophisticated programs are fairly complex to learn and master. However, costs can be saved by creating printed materials in house rather than paying outside vendors. Managers may want to consider desktop publishing skills when developing job descriptions for staff support positions.

Presentation Applications

Presentation graphics programs allow the user to create charts and graphs that consolidate data into visual formats. The final product is much more attractive than spreadsheet graphics. The lines are cleaner, there are wider choices of type styles and sizes, and the graphics can be produced in many colors.

Most presentation graphics programs create charts from a table of numbers that is entered directly into the program. Some programs can import a table that has already been created on a spreadsheet, thus saving time and labor.

Whether a health promotion department needs a presentation graphics program depends on the audience it wants to reach. It is highly desirable to use top-notch graphics in presentations to senior management and customers. As with desktop publishing, skills in the use of graphics applications should be considered for inclusion in the job descriptions for support staffers.

SPECIAL APPLICATIONS

There are software programs written specifically for use in health promotion environments. Four of these are health risk assessments, nutritional assessments, facility management, and management information systems.

Health Risk Assessments

Health risk assessments (HRAs) have become a core tool in health promotion. They provide a participant report with valuable data that can increase awareness, establish a baseline for physiological measurements, and often supply the motivation to change behavior. Often this data is converted into a health rating such as a score or age. For the program manager, HRAs provide baseline data for the participant population and tools for working with participants in health counseling and program direction. HRAs usually are administered in conjunction with a general health screening activity at the start of a new program and annually thereafter.

In selecting an HRA program, there are a number of factors to consider. The first is whether to process the HRA in house or have it done by a vendor. Factors that bear on this decision are the availability of suitable hardware and qualified employees and the cost of using a vendor. For in-house processing, the choices are to purchase software or license it from the distributor.

The next decision is what features the HRA program must have. Some common features are:

- Range of physical measurement options
- Length of questionnaire
- Age bias
- Range of questions, e.g., does it include mental health concerns?
- Reading level
- Output document quality
- Ease of interpretation
- Ability to interface with MIS software
- Method by which risk is portrayed, e.g., risk age vs. health score or other means

The selection of an HRA program is an important decision because the department will probably want to use the same instrument for several years to make comparisons. Before making this decision, managers should discuss their needs in detail with prospective vendors.

Nutritional Assessments

Nutritional assessment software programs are useful adjuncts to nutrition and weight management programs. They allow the user to enter a person's food intake and obtain an output document with nutritional content and suggestions for improvement. Some programs offer meal planning ideas and shopping lists. As with HRA programs, it is possible to process assessments in house.

In selecting a program, managers should consider their needs carefully. The needs of an educational and motivational program are considerably less sophisticated than those of a clinically based nutritional counseling program. Available software features include

- data input flexibility,
- length of time for which food intake can be logged and entered,
- educational value of the output document,
- age appropriateness, and
- meal planning and shopping list options.

As with most computer programs, there is a wide range of prices for nutritional assessment software. It is easy to spend more than necessary and purchase features that have no benefit in a user's specific application. Before making a decision, managers should discuss their needs thoroughly with prospective vendors.

Facility Management

Facility management programs are fairly sophisticated database management software used to track participant attendance and activity in a fitness facility. They also can be used to track dues and print invoices and financial reports.

Facility management programs usually are designed for commercial fitness club use and can be expensive. However, they are a valuable tool for managing and reporting facility activity. Consider these features:

- Ability to record workouts for a variety of activities
- Option of graphing workouts over time
- Ability to compare workout with exercise prescription, such as flagging workouts above and below target heart rate zone
- Incorporation of test battery results for time comparisons
- Calculation of caloric expenditure for workouts

These factors, as well as data that shows progress toward or completion of the program goals, should be considered in selecting a software package.

Management Information Systems

Management information systems (MIS) are powerful database tools for managing and reporting on an entire health promotion program. Some of the capabilities of an MIS are listed here:

- Extremely large record capacity
- Receiving employee data from the human resources department directly from tape or disk
- Accepting HRA data from disk
- Tracking fitness activity, class attendance, and any other activity through user-defined fields
- Tracking complex incentive/point systems
- Creating mailing labels, mailing lists, and mail-merge letters
- Selecting lists of participants by any criteria useful to program managers
- Creating custom and standard reports for any activity or group of activities

The primary difference between a packaged health promotion MIS program and a database management system is that the MIS software has been programmed specifically to meet the needs of health promotion. It is possible to program a database to meet health promotion needs, but the cost would be high compared with that of a prepackaged program. However, for less sophisticated health promotion programs, the simpler database approach may be appropriate.

Mini/Mainframe Applications

In some organizations, the human resources department stores data in a minicomputer or a mainframe computer. This data can be useful to health promotion managers. Also, many organizations have electronic communication systems that can be used to transmit information about health promotion programs.

Human Resources Data

Human resources departments must track large amounts of employee data to perform their functions and comply with state and federal laws. For health promotion programs, two useful kinds of data are demographics and health claims. For managers who have access to this data, it will be much easier to apply many of the marketing methods described in Part III.

For demographic information, health promotion managers may be able to obtain from the human resources department custom or standard reports with data on location, job type, age, sex, or any other factor that has been tracked.

Information about health claims may be less accessible. The human resources department may not maintain these data on its computer system. Claims may be tracked by an outside vendor, and the human resources department may be able to share only the reports it considers standard. Claims information also may be under the control of the employee benefits department or medical director. Custom reports may be costly and should be requested only after careful consideration.

Communication Systems

Many organizations have electronic communication systems such as electronic mail (E-mail) and electronic bulletin boards that can be useful to the health promotion department.

E-mail has become a standard communication tool for many organizations. It consists of a network of computers connected by either wire or telephone modem, and software that allows communication. Users are assigned "mailboxes" and can type notes and send them to other users. With a single keystroke one can send messages or even complete documents to large distribution lists across the country. The advantage is that messages can be directed to specific people instead of being sent to everyone.

Electronic bulletin boards are similar to E-mail except that, unlike an individual E-mail box, everyone on the system can look at an electronic bulletin board. The owner of a bulletin board can post information, and other users of the system can read it. The disadvantage is that only those interested in the bulletin board will look at it. To keep people interested, the owner needs to maintain it, just like a real bulletin board. Many systems also have an interactive bulletin board feature that allows users either to have limited posting rights or to respond easily to the owner. The organization's MIS department will have clear guidelines that determine ownership, maintenance, and access to bulletin boards.

Bulletin boards and E-mail are useful for communicating current program information, general health messages, and calendars. They can be used interactively to connect dispersed participants with similar interests such as a national running club event. They also can be used to exchange software and data files electronically by a process known as downloading.

SUMMARY

Computer applications allow health promotion managers to function more effectively and efficiently. Managers should become skilled in their use and knowledgeable about their applications. Standard software such as word processing and spreadsheets are a minimum requirement for managers. Database applications, desktop publishing, and presentation graphics move managers to a higher level of sophistication.

Software for special applications such as health risk assessments, nutritional assessments, facility management, and MIS tools should be considered in program planning. Managers also will find it easier to apply many of the marketing principles described in the next part if they have access to demographics and health claims data that may be maintained by the human resources department.

KEY TERMS

cells (p. 91)
electronic bulletin boards (p. 96)
electronic mail (p. 96)
fields (p. 92)
mail merge (p. 91)
MS-DOS (p. 89)
record (p. 92)
spreadsheet (p. 91)

SUGGESTED RESOURCES

Brownstein, M. (1989). *Using dBase IV: Basics for business*. New York: Wiley.

Gunton, T. (1990). *Inside information technology: A practical guide to management issues*. Englewood Cliffs, NJ: Prentice-Hall.

Keen, P. (1991). *Shaping the future: Business design through information technology*. Boston: Harvard Business School Press.

Knight, A., & Silk, D. (1990). *Managing information: Information systems for today's general manager*. New York: McGraw-Hill.

Schutzer, D. (1991). *Business decisions with computers: New trends in technology*. New York: Van Nostrand Reinhold.

Swartz, R. (1989). *Doing business with computers*. Englewood Cliffs, NJ: Prentice-Hall.

Part III

Marketing

In Part II we focused on the people and departments that make up an organization. In Part III we move away from the organization's relationships and travel into another functional area: marketing. Marketing is a process that directs the flow of goods and services to consumers. The goal is to have satisfied customers and to meet the organization's goals and objectives. We focus on specific techniques that will help managers direct health promotion services and products to consumers.

We begin our discussion with an overview of marketing in chapter 11. In chapter 12 we examine the customers for health promotion services. In chapter 13 we consider customers' demands for services vis à vis the organization's supply and that of its competitors. In chapter 14 we examine the factors that compete with health promotion activities in the marketplace. Chapter 15 begins an explanation of how to identify a market for health promotion services, including gathering data and determining specific market segments. In chapter 16 we use this information to develop a marketing strategy. We end Part III with suggestions for how to promote a program.

In our discussion of marketing we will address the issue of competition. For purposes of illustration we will consider competition from the standpoint of worksite health promotion. For some health promotion managers, worksite programs are the competition. However, the concepts we present apply to all health promotion situations. Of course, individual managers may need to look at the examples from a different perspective.

Many excellent products and services have been developed that few people have purchased. A likely explanation is the failure to employ effective marketing techniques. After completing Part III, health promotion professionals will understand the steps involved in developing a strong marketing program.

Chapter 11

The Open Market Analogy

As the field of health promotion has evolved, it has moved from a fitness center-based program staffed exclusively by exercise physiologists to more comprehensive programs staffed by health education generalists. Some programs have fitness centers at their core; others do not. Some are based in the human resources department; others are part of an employee benefits unit. Some emphasize prevention of cardiovascular diseases and cancer; others focus on actual health care claims problems.

The primary reason this evolution has yielded such a wide range of program configurations is that programs are no longer driven by what a "health expert" believes a population should be doing, but rather by what the population needs and wants. This is the way a market-driven service operates.

The most successful health promotion programs today are market driven. For this reason, it may be useful for managers to make an analogy with an open market situation, looking at their programs as if they were independent entrepreneurs. In this analogy, all the elements of a health promotion program can be compared conceptually to the open market. Each of these elements is influenced by market forces. Managers who operate their programs as a service business in the open market and use effective marketing principles and techniques usually will have more successful programs by virtually any standard. Using this analogy, we can draw a parallel with our definition of running a successful business:

> If a business can offer a product or service that meets the needs of a market

segment at a price they can afford at least as well as the competition, and one can tell them about it, the business will be successful.

MARKET ELEMENTS

A market system has a number of basic elements. It will be helpful to relate some of these elements to the various elements of a health promotion program. The market elements we will consider are

- products and services,
- suppliers,
- customers,
- prices,
- market research,
- advertising, and
- competition.

Products and Services

Products and services are bought and sold in the open market. Products are tangible goods that may be consumable, like food, or nonconsumable, like diamonds. Services are human actions that provide a benefit to the end user. In health promotion programs, there are a number of elements that can be viewed as products, including

- educational materials,
- incentive items such as T-shirts,
- health risk appraisal results booklets, and
- newsletters.

However, there are many more program elements that can be viewed as services. These include

- physical testing,
- classroom instruction,
- fitness center access,
- exercise prescription,
- health consultation,
- organizations such as running clubs, and
- motivational activities.

Because of its many service-related elements, we will consider health promotion a service business for purposes of the open market analogy and the subsequent discussions of marketing.

Suppliers

Any person or organization that provides a product or a service can be considered a supplier. In the field of health promotion, suppliers include the following:

- Health promotion staff members
- Related departments, functions, or employees in the organization, such as the medical director or safety manager, that are cooperatively providing products and services
- Vendors of health risk appraisal tools
- Vendors of health screening services
- Vendors of newsletters
- Printers and graphics vendors
- Organizations that manage fitness centers
- Contracted aerobics instructors

Customers

Customers are the end users or consumers of products and services. They make the decision to purchase or not purchase products and services, based on their needs and desires and on the quality and prices of the items being offered. In a health promotion program, the customers may include

- the organization that is funding the program,
- the participants or employees, and
- the dependents of the participants or employees.

Prices

A price is the quantity of one thing that is exchanged or demanded in barter or sale for another thing. In the open market, prices typically are set and paid in money; less commonly, goods or services may be exchanged. In a health promotion setting, participants may have to purchase products or services unless program costs are paid by the sponsoring organization. The price of participation may manifest itself as less tangible "currency," such as personal time and inconvenience.

Market Research

Market research is the process of gathering data on the needs and characteristics of a market and the nature of the competition. In a health promotion setting, market research can be conducted by means of

- employee surveys,
- community resource surveys,
- wellness committees,
- human resources reports, and
- information in the public domain.

Market research is discussed in detail in chapter 15 and in relation to long-range planning in chapter 21.

Advertising

Advertising is a primary means of informing prospective customers about a product or service. It is useful in developing brand name recognition and influencing buying decisions. It is typically done in print media such as magazines, journals, and newspapers, through direct mail campaigns, or on radio or television. In a worksite health promotion program, advertising usually is done via

- flyers,
- posters,
- paycheck stuffers, and
- mailings to employees' homes.

Advertising is discussed in detail in chapter 17.

Competition

Competition refers to products and services a customer can choose from other providers. A worksite health promotion program may be competing with these options:

- YMCA and YWCA programs
- Private health clubs
- Commercial weight control, stress management, and smoking cessation programs
- Medically based programs offered by physicians, health maintenance organizations, and hospitals
- Personal trainers
- Personal fitness programs

Other Competition

Here are other situations which may result in competition with health promotion programs:

- Family time
- Car pools
- Company-sponsored recreational events such as bowling leagues
- Work-related situations such as overtime
- Television
- Other recreational pursuits

Competition is discussed in more detail in chapter 14.

MARKET FORCES

If one views a health promotion program as an independent business venture operating in an open market, it follows that many of the same forces found in the open market will influence the program's operation and decisions made about it. We already have examined a number of market forces individually. We now will consider how market forces interact to influence supply and demand.

Supply and Demand

Supply and demand refers to the dynamic relationships between the availability of products and services and the amount that consumers want at any point in time. In health promotion, supply is represented by these elements:

- Number of classes offered
- Availability of appointments for fitness tests or health risk assessments
- Availability of a health professional's time for individual consultation
- Frequency of a newsletter's publication
- Number of competing programs available

Demand is represented by these elements:

- Number of employees in a given risk category
- Number of employees who are interested in the program's services
- Management decisions that mandate participation in any aspect of a program

Health promotion managers can control some of the forces that affect supply and demand for their services; other factors cannot be controlled. Awareness of the relationships between these market forces is essential in planning and evaluating programs. Supply and demand are explored further in chapter 13.

APPLICATION OF MARKETING PRINCIPLES AND TECHNIQUES

As managers adopt an entrepreneurial mind-set, they find it productive to consider several factors in planning to market their health promotion programs. These factors often are described as the Ps of marketing. The number of Ps that are listed will vary depending on whose opinion is asked. For the purposes of health promotion, one can easily identify six Ps that facilitate discussion:

- Product
- Price
- Place
- Promotion
- Packaging
- Position

A brief description of each factor is given in Table 11.1. We discuss each of these factors in a variety of contexts throughout the remainder of Part III. Managers will see that the factors are interrelated and often overlapping.

SUMMARY

Health promotion managers will find it useful to adopt the mindset of an entrepreneur who must anticipate, plan for, and respond to all of the forces that operate in the open marketplace. Treating health promotion activities as products and services that compete for customers' dollars helps managers make sound decisions about how to operate their programs.

To develop a complete market-driven program, managers must address the factors of product, price, place, promotion, packaging, and position.

Table 11.1
The Six Ps of Marketing

Product	Actual product or service offered
Price	Amount of value given in exchange for products and services
Place	Specific purchasers of products and services
Promotion	Informing the market about the products and services
Packaging	Appearance of products and services
Position	Portrayal of products and services to the market in relation to its features and benefits

Adapted from the traditional four Ps of marketing.

To succeed, a business must identify and meet the needs of its customers. In the next chapter we explore specific techniques that are effective in marketing health promotion services.

KEY TERMS

customer (p. 102)
price (p. 103)
products (p. 102)
services (p. 102)
supply and demand (p. 104)

SUGGESTED RESOURCES

Bernstein, A. (1988). *The health professional's marketing handbook*. Chicago: Year Book Medical.

Cooper, P. (Ed.) (1985). *Health care marketing: Issues and trends*. Rockville, MD: Aspen Systems.

Cooper, P. (Ed.) (1986). *Responding to the challenge: Health care marketing comes of age*. Chicago: American Marketing Association.

Hillestad, S., & Berkowitz, E. (1991). *Health care marketing plans*. Gaithersburg, MD: Aspen.

Manoff, R. (1985). *Social marketing: New imperative for public health*. New York: Praeger.

Self, D., & Busbin, J. (Eds.) (1990). *Marketing for health and wellness programs*. New York: Haworth Press.

Sweeney, R., Berl, R., & Winston, W. (Eds.) (1989). *Cases and select readings in health care marketing*. New York: Haworth Press.

Chapter 12

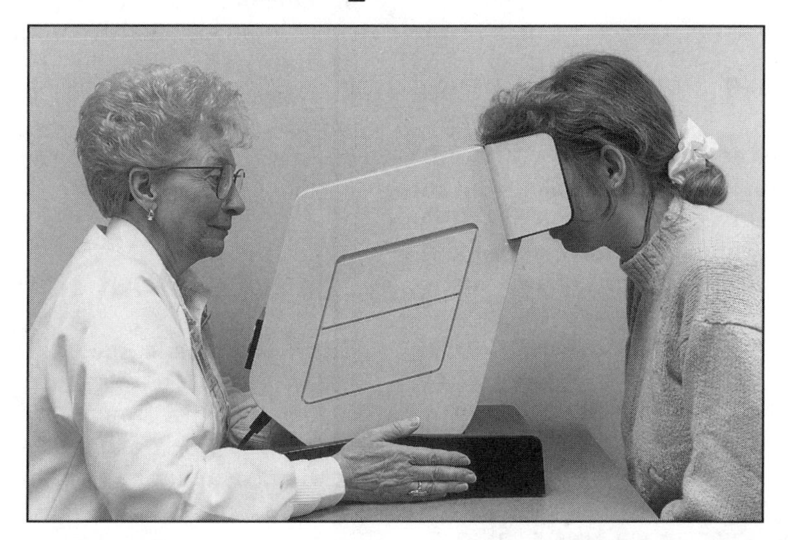

Consumers of Health Promotion Services

As organizations institute health promotion programs, two customer classes are created: the organization itself and its employees and their dependents. Fortunately, the needs of these groups are synergistic. That is, if the program meets the needs of employees and dependents, usually by assisting them in improving their health and lifestyle, the organization will experience improved productivity and morale and reduced health care claims. Everybody wins.

Understanding the characteristics and dynamics of the consumers of health promotion services is critical to the success of a program. In this chapter we focus on the end users of health promotion services, their needs, and some relevant issues.

EMPLOYEES AS END USERS

The term *end user* is applied to the person or entity that is intended to be the ultimate consumer of a product or service. Because employees are the consumers of health promotion services, they become the primary targets of program interventions. Their success also will yield success for the organization. As the end users of health promotion services, employees are the market that must be analyzed and whose needs must be met. This analysis will identify specific market segments, influence program content and direction, suggest promotion and communication strategies, and provide other information necessary for the program's success. This information will serve as input to the development of marketing and planning activities.

DEPENDENTS AS END USERS

As we examine the market for a health promotion program, it will become clear that in most organizations there is a group that may be at least as important as the employees due to their numbers and the impact of their health care claims. This group is the dependents of employees. When dependents receive health benefits from an organization, the organization has the same concerns about their health as it has about the health of employees. If dependents are in high-risk categories, their claims experience will be higher and the organization's costs will rise.

Like any subgroup, dependents present their own set of needs, restrictions, and obstacles to participation in health promotion programs. Managers must subject dependents to the same marketing analysis as the employee group and design program activities to meet their needs.

CONSUMER NEEDS

At the heart of any market-driven venture are consumer needs. In the area of health promotion, several variables affect the consumer's ability to participate and willingness to purchase services.

- Health risks
- Personal interests
- Learning style
- Flexibility of personal time
- Transportation
- Reading level
- Income
- Social needs

Although it may be impossible to take all of these factors into account, the more that can be addressed, the greater the chances of high program participation. Remember that in most health promotion environments, participation is optional. To meet the organization's goals for participation and related benefits, managers must apply marketing principles wisely and meet the needs of the program's consumers.

CONSUMER DECISIONS

There are several factors besides price (p. 103) that affect a consumer's purchasing decisions in health promotion programs. These factors are program effectiveness, quality of service, customer loyalty, and customer purchasing habits.

Program Effectiveness

The key to the long-term success of any business is having satisfied customers who not only return for additional purchases, but also provide the most effective form of advertising: positive word-of-mouth referrals. In health promotion programs, the key to customer satisfaction is program effectiveness as measured by success in meeting customer needs and expectations; simply put, a program has to work. With professional guidance and motivation, and with commitment and reasonable effort on their part, participants should be able to successfully quit smoking, lose weight, improve cardiovascular fitness, or otherwise achieve their health objectives.

Quality

Quality is closely related to customer satisfaction. Most modern definitions of quality relate it to customer expectations. Meeting or exceeding those expectations is one measure of quality. Meeting them consistently is another measure. Having a quality program requires that you understand what the expectations of your customers are and develop programs to meet their needs, wants, and expectations. Meeting them consistently requires that you have detailed and documented processes in place so that your program can be replicated at different times and places. The subject of quality goes beyond this text and will not be addressed in detail. Most large organizations will undoubtedly have formal quality training programs in place.

Customer Service

The decade of the 1990s is proving to be an era of competition for consumers that is marked by efforts to provide a high level of customer service. Providing excellent customer service is easy and costs nothing. It begins within the organization, with a customer being defined as anyone who receives the product of one's work. The manager is the customer of anyone who submits a report. The employee who submits a report to be typed is the customer of the person who types it. If employees are encouraged to follow the Golden Rule in their dealings with each other, they will learn to behave that way toward customers.

In health promotion programs, both professional and interpersonal skills are essential elements of good customer service. Employees should be trained to deal with participants in a courteous, positive, helpful way.

Managers can monitor service by setting standards that are easily measurable. Here are some examples of customer service standards:

- Response time for picking up the phone (for example, no more than three rings)
- Turnaround time for information requests
- Waiting time for health consultations
- Number of complaints for a defined time period

Customer Loyalty

Consumers are creatures of habit. Once they are satisfied with a service, they are reluctant to switch to another supplier unless they become dissatisfied or the competitor can lure them away with "a better mousetrap." Building loyalty should be a fundamental goal of a health promotion program. The initial level of customer satisfaction is the foundation on which loyalty is built. Managers must guard against becoming complacent and losing touch with customer needs and attitudes. Developing responsive employees and striving to provide higher levels of service help build and maintain customer loyalty.

Because services like health promotion programs are intangible, and once delivered tend to be forgotten, customers need regular reminders of the service they have received and of their satisfaction with it. In a health promotion program, these reminders could include the following:

- Regular reports to management on program successes
- Individual reminders for blood pressure checks
- Weight maintenance follow-up mailing to participants who have completed an initial weight loss program
- Thank-you memos or cards to employees who have participated in a program activity

The consumers of health promotion programs need a steady diet of positive information about the program. This will increase the likelihood that they will remain loyal to the program despite the efforts of competitors.

Purchasing Habits

Within a group of consumers of a product or service, individual consumers base their buying decisions on different factors. It is useful to understand purchasing habits when planning the promotional aspects of a program. These are some of the common purchasing habits:

- Impulse buying
- Comparison shopping
- Brand name shopping
- Bargain hunting
- Fad buying

Understanding these habits as they relate to health promotion participants can make the "sale" easier.

Impulse Buying

Impulse buyers tend to make purchasing decisions on the spot and without much thought. They usually focus on the features of a product or service rather than on the benefit. They are likely to fall prey to advertising gimmicks, sales, and special offers. Many of their decisions are poor. Impulse buyers will purchase home exercise equipment that promises a firm tummy with no work. They will use diet pills to achieve quick weight loss rather than pursue the long-term health benefits of a sound program.

Impulse buyers often can be enrolled in program activities during short-term special promotional events in cafeterias. They also tend to be responsive to peer pressure. It is helpful to make the registration procedure simple and quick for this group. Clever incentive items also can be useful.

Comparison Shopping

Consumers who comparison shop consider many factors before making their decisions. They do extensive research using independent consumer reports, manufacturers' brochures, and other data. They consider features, benefits, and price. They tend to be prudent and wise consumers who are not easily swayed by gimmicks.

In health promotion, comparison shoppers should be provided scientific data that support both the need for a change in health habits and the method offered by the program. Use of health risk assessment tools with detailed report options can be helpful in persuading these buyers to participate. To accommodate their long decision-making process, these buyers need plenty of lead time before a registration deadline.

Brand Name Shopping

This group develops strong brand name loyalty and can become a significant customer base. They respond to the psychological impact of repetitive advertising in developing name recognition. All other factors being equal, they will base their decisions on their familiarity with a brand name and the associations they make with that brand.

In health promotion, brand name shoppers are likely to be attracted to programs that use particular makes of equipment with which they are familiar, or health care products they have used and liked. Employees in this group may have to be dislodged with the use of clever program names from their loyalty to competing "brands" such as a local health club or gym. Once brand name shoppers are committed to the health promotion program, however, they are likely to remain loyal to it for some time.

Bargain Hunting

Bargain hunters like to believe they are getting the best deal possible when they make a purchase. Whether they actually did get the best deal may not be as important as the fact that they believe that they did. In the retail world, good deals are exemplified by discounts, sales, incentives, limited-time offers, and negotiated terms. In the process of searching for the ultimate deal, bargain hunters may appear to be performing the same thorough research that a comparison shopper does. Bargain hunters, however, are not as meticulous and may pursue savings at the expense of real value.

In a health promotion setting, bargain hunters may respond to the pressure of limited enrollments, "early-bird" registration discounts, or incentive items for the first 30 people who sign up for an event. They may select an inferior program because of lower price, without looking closely at the quality of the program. They may opt for a quick weight loss program that promises permanent success rather than a slower, more medically sound program.

The key to selling bargain hunters on a health promotion activity is to give them the feeling that they got the "best deal," however that is defined. That may mean selling them on the value of the program. Low recidivism rates that result in less expenditure of time and money over the long term may be the best deal. Employee copayment rebates after successful lifestyle changes may be the incentive that results in a lower net cost and constitutes a "good deal." Remember that although offering program activities at the worksite at no cost may be viewed as a "good deal" to a bargain hunter, if something is absolutely free, it may be assumed that it has no value. If the organization's policy is to subsidize programs 100%, once again, the value must be amplified to the participants to offset this negative assumption.

Fad Buying

Fad buyers tend to be strong conformists. Although they may evaluate prospective purchases, if everything else is equal, they usually go with the popular choice. Like impulse buyers, they are not thorough shoppers. Their choices may be based not on quality or value but on color or style. Their motivation for buying is their desire to belong.

In health promotion, fad buyers are attracted to programs that have a strong social orientation, such as group weight loss activities or contests. They may sign up for a program just to earn an incentive their friends have that they cannot get any other way. Often they can be enrolled if their entire social group is approached. Mentioning the names of popular or admired fellow employees who are participating in a program also may induce fad buyers to enroll.

CONSUMER EDUCATION

Potential consumers of health promotion services may lack information that could help them make decisions. They may not know they have needs that can be met by these services, or they may not have enough information to make intelligent purchasing decisions. These needs for information can be filled with consumer education.

In the process of educating consumers, health managers should begin with a broad consciousness-raising perspective rather than focusing on the specifics of an individual activity. For example, rather than being told about the benefits of regular blood pressure monitoring, a consumer may need more information about hypertension as a disease, who can be affected, and the causes and consequences. People who find the information relevant will identify themselves as candidates for more specific information and ultimately for interventions.

Although this self-selection process can be valuable in reaching target populations before activity-specific promotion begins, it may be useful to offer a second level of education before moving to the activity promotion. This level can be aimed at those whose interest was piqued by the general information but who are not quite ready to commit themselves to an activity. This second-tier approach can be as simple as continuing to raise awareness through additional informational materials or as complex as an activity specifically designed to intensify the latent anxiety of people who are at marginal risk levels.

As part of a health promotion program's overall communication plan, it is effective to provide general education about a wide range of topics that lead into the program's activities. These topics can be the fodder for newsletter articles, topical campaigns, and pamphlets. "Piggybacking" on current events and public health agency campaigns such as the American Heart Association's efforts during Heart Month can provide additional structure as well as themes and educational materials for the program.

SUMMARY

Using the open market analogy, it is logical to regard the participants in health promotion programs as consumers of health promotion services. Knowing that both employees and their dependents are the end users helps managers design programs that meet users' needs and that appeal to their motives as buyers. Participants may need several levels of education about health issues before they are ready to make decisions about participating in specific program activities.

Another important factor in marketing operations is the relationship between supply and demand, which is a central force in open markets for products and services and affects both competition and price. In the next chapter we begin an overview of supply and demand.

KEY TERMS

consumers (p. 107)
end users (p. 107)
quality (p. 109)

SUGGESTED RESOURCES

Ambry, M. (1990). *The almanac of consumer markets: A demographic guide to finding today's complex and hard-to-reach customers*. Chicago: Probus.

Cambridge Reports. (1986). *The new consumer. 51 opinion and behavior trends shaping today's marketplace*. Cambridge, MA: Author.

Evans, J., & Lindsay, W. (1993). *The management and control of quality*. Minneapolis: West.

Lazer, W. (1987). *Handbook of demographics for marketing and advertising: Sources and trends on the U.S. consumer*. Lexington, MA: Lexington Books.

Olson, J., & Sentis, K. (Eds.) (1986). *Advertising and consumer psychology*. New York: Praeger.

Shilliff, K., & Motiska, P. (1992). *The team approach to quality*. Milwaukee: ASQC Quality Press.

Sirgy, M. (1984). *Marketing as social behavior: A general systems theory*. New York: Praeger.

Tomlinson, A. (1990). *Consumption, identity, and style: Marketing, meanings, and the packaging of pleasure*. New York: Routledge.

Wickam, P. (Ed.) (1988). *The insider's guide to demographic know-how: Everything marketers need to know about how to find, analyze, and use information about their consumers*. Ithaca, NY: American Demographics Press.

Chapter 13

Supply and Demand

One of the most complex, influential, and dynamic forces in the open market is the interplay of supply and demand. Supply and demand are subject to both internal and external influences. Understanding how these forces affect each other will help health promotion managers develop an effective marketing approach (see Figure 13.1). Knowledge of supply and demand also will help managers prepare more accurate forecasts and will allow them to control some of the factors involved.

INFLUENCES ON DEMAND

The demand for health promotion services is affected by factors both within and outside an organization.

Internal Influences

At the worksite, internal factors include policy changes, employee deaths or illnesses, reorganization with new management, work cycles including overtime, and shifts in the demographics.

Policy Changes

One of the most significant internal influences on the demand for health promotion services is changes in organizational policies, particularly those related to health behavior and employee benefits. For example, when management implements a smoke-free workplace policy, there is likely to be an increase in the number of smokers who want to participate in smoking cessation classes.

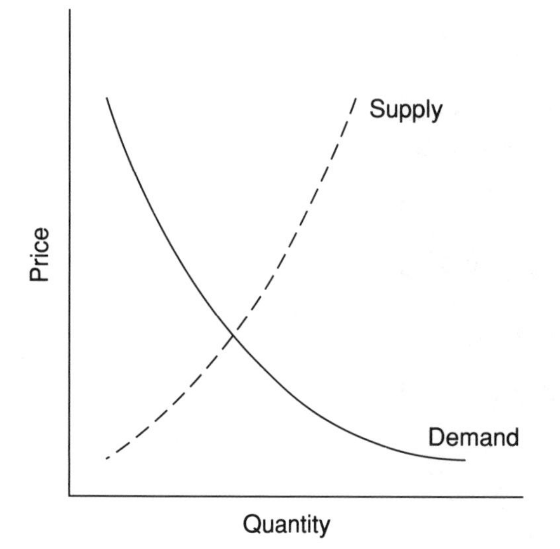

Figure 13.1 The relationship between supply and demand when compared with price and quality. Seasonal variations affect participation in activities.

A second policy that may affect demand involves incentives for seat belt usage. Management may offer an additional life insurance benefit if an employee dies in an automobile accident while wearing a seat belt. Management may request a seat belt awareness campaign to coincide with the announcement of the policy, and to be conducted periodically thereafter.

A third policy with important implications for demand is one that mandates drug testing or a drug-free workplace. In cooperation with the employee assistance program, the health promotion department may be asked to assist in conducting drug awareness classes and coordinating the overall communication effort.

Also affecting the demand for health promotion services are policies under which the organization encourages program participation by means of copayments, payroll deductions, and similar mechanisms. To a certain point, such mechanisms can be expected to increase demand.

Demand also is influenced by policies with respect to working hours. If an organization operates on flex time, which gives employees some latitude in choosing their hours of work, it is likely that the demand for health promotion services will increase.

Other policies that may affect demand include the following:

- Facility improvement policies that assure showers be included in lavatory remodeling or new facility specifications.
- Policies that restrict the use of areas such as cafeterias for aerobics classes.
- Policies that require employees to sign waivers or disclaimers before participating in health promotion activities.
- Policies that dictate the vendor selection process and restrict the use of vendors for portions of the program. This might make it difficult to find qualified vendors.
- Policies that restrict or prohibit activities that might be construed to be medical in nature, such as fitness testing or blood pressure monitoring.

Employee Death or Illness

A powerful internal factor that affects demand for health promotion services is the death or serious illness of an employee. This is especially true if the employee has high visibility in the organization, and if the illness or cause of death is clearly a lifestyle-related disease such as lung cancer or heart disease. Many employees may examine their own lifestyles and begin to grasp the importance of health promotion messages that have been communicated in the past. Some employees may show increased interest in activities related to the risks associated with the stricken employee's illness. Others may pursue activities related to their own health risks or become interested in improving overall health and fitness.

Reorganization With New Management

Even when a health promotion program is well established and has a long-range plan approved by senior management, a reorganization with new management is bound to affect demand. Whether demand will increase, decrease, or remain relatively stable will depend on how the new managers view health and fitness, both from a personal perspective and from the standpoint of its value as an employee benefit.

Reorganization also may result from merger, acquisition, or divestiture. Even if management philosophy remains stable, there is likely to be significant fluctuation in populations where new groups are added or existing groups are removed. These changes can be expected to cause some shifts in demand for health promotion services.

Work Cycles

Work cycles can affect demand in both highly predictable and highly unpredictable ways. The predictable cycles are peak work periods, such as income tax season for an accounting firm or the pre-Christmas period for a department store chain.

During peak work periods, affected employees may decrease their demand for activities that require significant time commitments. In contrast, there may be stronger demand for stress management activities or activities with minimal time commitments. During the off peak season, demand for activities is likely to return to normal levels.

There are some unpredictable work cycles that can affect demand for health promotion activities. One is the unexpected scheduling of overtime work, which can cause a temporary decline in demand. Layoffs or the rumor of layoffs can have a similar negative effect on demand for program activities. When employees fear the loss of their jobs, they move down Maslow's hierarchy of needs (Maslow, 1970) and become concerned with survival rather than quality of life issues.

To minimize rumors and to maintain employee morale, and possibly to avoid alarming stockholders, senior managers in most organizations plan layoffs and major changes in work cycles in an atmosphere of secrecy. Health promotion managers typically are not involved in these early planning stages. However, by building and maintaining strong communication with senior managers, a health promotion manager may be able to obtain advance information about work cycle changes or layoffs.

Demographic Shifts

Because a health promotion program's long-range plan is based on a number of assumptions, including employee demographics, any significant changes in those assumptions will affect the plan. The changes in demographics that will affect demand include:

- Increases or decreases in the general population.
- Changes in the population of any segment to which programs have been targeted, for example, gender, education level, and job type, such as clerical or manufacturing.

External Influences

The external factors that can affect demand for health promotion programs include seasonal influences, the media, medical research, current public health issues, and market forces, including competition.

Seasonal Influences

Seasonal influences are fairly predictable and are routinely considered in health promotion planning. Seasonal factors may affect both overall demand and the demand for specific programs. Some of the more common effects are listed below:

- Increased interest in January in weight control and other activities based on New Year's resolutions
- Increased interest in fitness-related activities toward the end of winter, especially in the north
- Increased interest in weight control activity in preparation for winter vacations in warmer climates and for the beginning of summer
- Decreased willingness to participate in indoor after-work activities in the late spring and summer
- Decreased interest in vigorous outdoor activities in extremely hot and/or humid climates
- Increased interest in local seasonal sports like outdoor running or cross-country skiing

More specific seasonal influences can be seen in relation to holidays, both traditional and "promotional." Examples are:

- Increased interest in stress management activity over the November and December holidays
- Increased interest in activities connected with Heart Month, Cancer Month, National Jogging Day, and other special-interest efforts

Anticipating the effects of these predictable shifts in demand will allow for better program planning and demand forecasting.

Media Influences

The media have become a powerful influence in the lives of Americans. Advertising, news coverage, Hollywood fantasies, and other forces strongly affect people's perceptions of what is healthy and what is not. In many cases, such as the publicity surrounding cholesterol, the influence has been positive. In others, such as the "American ideal" of feminine beauty, it has not.

In any event, the power of the media can influence demand in health promotion programs. Managers should be aware of media influence and should know how to use the momentum created by the media to enhance their programs.

Here are some examples of media influence:

- Public service campaigns such as "Sober Cab" programs, seat belt compliance, and anti-smoking messages
- Publicity in conjunction with designated health periods as outlined earlier
- Publicity about health problems of highly visible public figures such as Rock Hudson's death from AIDS, Betty Ford's recovery from alcoholism, and the health challenges of some professional athletes
- Feature articles in national magazines that open the door for participant inquiries about specific topics
- Fad exercise or diet programs that may need counter-publicity by the program

The challenge of media influence is that it is largely unpredictable. The death of a movie star cannot be anticipated. By keeping current with media treatment of health issues, managers can provide follow-up education to address short-term concerns of the population.

Research Findings

Significant health-related research findings may be reported accurately only in professional and scientific journals. The consumer media often focus on the aspects of a research project that have popular appeal rather than reporting all of the findings or even putting them in an appropriate context. Given the amount of health-related research in progress, managers should keep abreast of developments in core areas of concern to health promotion, primarily lifestyle-related issues. As new findings are released, demand for information tends to increase. Examples include the following:

- Reports that consuming foods like oat bran can reduce the risk of cancer and heart disease
- Reports about relationships between stress and disease
- Results of drug trials that appear promising but are inconclusive because they are preliminary
- Research in rapidly changing areas like cholesterol, where today's truths may become tomorrow's myths

Health promotion managers can serve their populations effectively by reviewing findings in core areas and reinforcing the research that seems to have merit. This can be achieved by means of special-interest seminars and newsletter articles written by health promotion professionals. Managers may decide to position the health promotion program as a clearinghouse for the latest information on areas of concern to participants. This can be done proactively with literature, seminars, and newsletter articles as well as by encouraging employees to use the staff as a resource and responding to individual questions.

Public Health Issues

Another factor that influences demand is reporting by the media on public health issues. Examples are AIDS, premenstrual syndrome (PMS), menopause, alcohol-related automobile deaths, and low-birthweight babies.

Whenever a public health issue receives media attention and captures the public interest, demand for information rises. How a health promotion program meets this demand will depend on the local level of interest, flexibility of the program schedule, and available resources. Staff members can respond to requests for information by distributing appropriate literature, publishing newsletter articles, arranging seminars, or conducting educational programs such as lecture series, holiday drinking/safety campaigns, or other activities.

True public health issues should be considered for inclusion in the overall program plan for the year. Short-term responses to increased demand may be needed when events, research, or media attention cause an increase in public awareness and curiosity.

Competition

There are many kinds of competition (as discussed in chapter 14), and any of them can affect demand for health promotion services. Most commonly, if there is increased competition for employees' time and interest, or if employees can have their needs met outside the health promotion program, demand for the program's services is likely to decrease. For example, if a local health club offers benefits that the organization's fitness center cannot match, participants will be drawn away from the program and demand will decrease. Alternatively, if the competing club advertises heavily, general interest in fitness may increase, and the organization's fitness center will have an opportunity to "sell" its benefits to a newly receptive population.

INFLUENCES ON SUPPLY

Like demand, supply can be affected by a number of factors: budgets, the internal political climate, labor shortages, and competition.

Budgets

The amount of funding a health promotion program receives has a direct impact on the supply of services it can make available. The budgeting process is discussed in detail in Part IV, Financial Operations.

When budget cuts force reductions in health promotion programs, managers can try to find alternative activities, such as free community agency programs, or funding, such as additional participant fees, or even refer participants to outside resources such as program competitors. When a program must cut services because of funding problems, the program may lose more momentum as employees sense a decrease in management's commitment to health promotion.

Internal Political Climate

The internal political climate of an organization can have a profound influence on the supply of health promotion services to employees. Political considerations affect budget decisions as well as attitudes that support or discourage program participation. The political climate may shift because of a change in ownership and/or management, or in response to changing business conditions.

Ideally, a health promotion program would be so firmly entrenched in the worksite culture that these shifts in support would be inconsequential. In reality, however, health promotion is one of many activities offered at the worksite that are considered discretionary.

Labor Shortages

Although a shortage of health promotion professionals is not a common occurrence, it can reduce the supply of programming. A manager may have difficulty filling positions in specific areas of expertise, such as aerobics or stress management instructors. Staff shortages in any area will force cancellation of classes and reduce the supply of services.

The degree to which a shortage of qualified professionals is a problem will depend on the size of the community, proximity to colleges and universities, availability of hospital-based community-outreach programs, and prevalence of health-related services in the area. Managers should research the local labor market before promising programs. Such research can form the basis for decisions about contracting for services, hiring new employees, using full-time versus part-time employees, and conducting internal training and/or certification programs.

Competition

Competition can be one of the most severe and uncontrollable factors that affect the supply of health promotion services to an organization's population. As competition increases from community programs or health clubs, the total supply of health promotion services increases. This has the effect of reducing demand for any given option. The challenge managers of worksite programs face is how to respond to competition.

The activities of competitors are unpredictable in both timing and nature. Health clubs can launch attractive membership drives at any time. Weight control programs can run promotional specials. In considering possible responses to competition, there are a number of options:

- Determine the timing of predictable competition such as local community education programs and schedule registrations to end before the competition's. Once they have signed up for a worksite program, participants are less likely to enroll in a community program.

- Tie all activities into an overall incentive program, such as an ongoing health point system, that "hooks" participants into a longer-range incentive, such as making significant selections from a point redemption catalog. Again, participants will be less susceptible to the lure of the competition.
- Stay in touch with employee needs and be flexible in meeting those needs.
- Be price competitive, remembering that the price that an individual pays is not only money, but also time and convenience.
- Track the activity of the competition. Get on the competitors' mailing lists and watch local newspapers. If they are offering special programs or packages, counter them with better offers to employees.
- Work to develop "brand loyalty" so that competition will not be as attractive on a subjective level. Participants who are loyal will not be likely to shop for alternative services.

FORECASTING

Supply and demand can be forecasted to a large extent if enough data are available. This is an important factor in project planning and budgeting.

Relationship to Project Planning

The nature of project planning, which is described in chapter 22, is that even the most carefully developed plan may need to be changed in response to changing circumstances. In a project plan, resources such as time, money, and labor are allocated according to project needs. The manager estimates the probable level of demand and plans to create a supply that will satisfy the demand. As the forecasted circumstances change, the plan may have to be changed to increase or decrease supply and resources.

Because of the dynamic nature of supply and demand and their effect on resources, accurate forecasting and tracking of demand are essential. If managers monitor closely the factors that affect supply and demand, they can implement adjustments to the plan in an efficient and timely manner.

Relationship to the Budget Process

As can be seen in the relationship between supply and demand and the dynamic project plan, a core item affected by changes is the budget. Understanding demand will enable managers to develop an accurate budget that allows the program to supply services of the kind and quantity demanded.

CONTROLLING DEMAND

Despite the volatile nature of supply and demand and the difficulty of anticipating and adjusting to changes in it, there are several techniques managers can use to help control demand and thereby reduce surprises in program management. Three of these techniques are increasing supplies of services, rationing services, and promotion.

In general, it is better to have some pent-up demand for services than to have an oversupply. When there is an oversupply of services, activities are conducted with partially filled rosters and are cancelled because of inadequate registrations. This conveys the impression that the program is not successful and hurts its reputation among employees. In contrast, if there is even a modest amount of pent-up demand, classes are filled and the existence of waiting lists gives the impression of popularity and creates a sense of urgency in the registration process.

Increasing Supply

As noted earlier, an increase in supply usually satisfies demand and may temporarily reduce it. There may be times when increased demand for a specific service forces managers to deviate from their plan. If it becomes too great a strain on resources managers may decide to cool the demand in a way that allows the program to get back on track.

For example, when extensive media coverage stimulates interest in a specific activity such as in-line skating, marathon running, or biking, requests for information may flood the health promotion department and all current activities may be filled. If the manager increases supply by offering additional sessions of existing programs, increasing points of distribution for materials, and writing additional newsletter articles on the subject, interest can be satisfied without a major disruption in the overall plan.

If the interest is serious and ongoing, managers must examine the options as they relate to the long-range plan and decide whether more significant program modifications are needed. If the interest cools in response to the additional supply of programming, the program can resume its course.

Rationing of Services

Rationing of services is one of the most useful techniques for regulating demand and bringing a program's schedule under control. It is particularly useful when a new program is introduced. Often during the initial stages of a program, because of intense promotion, participants are eager to sign up for every activity. Health promotion professionals tend to respond by striking while the iron is hot and scheduling additional sessions of popular activities. This often produces a negative result. Because of normal attrition, participation in the added sessions may decline to the point where discussion and small-group activities are hindered.

Another problem arises when, for example, high demand in the first quarter of the year causes the staff to schedule extra sessions in the second quarter in anticipation of continued high demand. During registration, participants tend to spread themselves out over all of the sections. The result is a number of either empty or partially filled classes. If some classes have to be cancelled, participants receive the impression that the activity is not popular. In addition, there are the administrative problems of notification, transferring, and rescheduling.

A more effective way of handling these high-demand periods is to stick to the plan and develop prioritized waiting lists. These convey the impression that the activity has value and create a sense of urgency among potential participants. Attrition decreases because a place in the classes takes on more value as a result of its apparent scarcity.

The ideal approach is to create a balance among meeting the needs of the population, following the plan, and maintaining demand through rationing. In this process, communication with employees is essential. Deadlines for registration and class limits should be clearly stated. Rules should be applied consistently to all participants, including executives. Follow-up and notification to people on the waiting list are also important.

Promotion

In the field of health promotion, promoting activities is often more difficult than providing the actual health services. Organizing accurate, current, and interesting health education information is not difficult. The challenge is motivating high-risk individuals, who by definition are resistant to health education and especially lifestyle changes, to attend activities. The primary purpose of promotion is to increase demand, and this is an ongoing challenge in all health promotion programs.

The most predictable scenario in a health promotion program, especially after it has been operating for a few years, is that of low demand. Managers may find themselves constantly trying to bring new life to old programs and revive awareness among the population. There is a tendency to believe that the population has been saturated with certain aspects of programming after numerous attempts to conduct ongoing activities have failed. In fact, it is highly unlikely, given what is known about the risk levels of most populations, that they have been saturated. It is more likely that they are bored with the approach and that it may need revision.

Another possibility is that the market segment that has been targeted needs to be changed. For example, if a fitness class for employees over age 40 has been experiencing declining participation, the health promotion staff may need to focus on a different risk area and target the class to another market segment.

Periods of low demand are opportunities to apply creative promotional techniques. Taking a cue from commercial advertisers' strategies—bold messages, discounts, and rebates—managers can increase participation in health promotion programs. Repetitive advertising, clever themes, special incentives, repositioning activities, and similar methods are effective ways to stimulate demand.

SUMMARY

Supply and demand are the most fundamental forces in the open market. Understanding their relationship is a key to success in managing health promotion programs. Both supply and demand can be influenced by internal and external factors, from budget considerations and organizational politics to media coverage and competitive forces.

Accurately forecasting supply and demand is important in project planning and budgeting. By understanding the dynamics of supply and demand and the factors that affect them, managers can exercise some control over demand.

In the next chapter, we investigate further the supply side of the equation. Specifically, we consider competition, including who and what may compete for health promotion program participants.

KEY TERMS

demand (p. 113)

supply (p. 117)

SUGGESTED RESOURCES

Adler, D. (1984). *Prices go up, prices go down: The laws of supply and demand*. New York: F. Watts.

Amara, R. (1988). *Looking ahead at American health care*. Washington, DC: McGraw-Hill.

Berliner, H. (1987). *Strategic factors in U.S. health care*. Boulder, CO: Westview Press.

Maslow, A. (1954). *Motivation and personality*. New York: Harper & Row.

Montgomery, D., Johnson, L., & Gardiner, J. (1990). *Forecasting and time series analysis*. New York: McGraw-Hill.

Saaty, T., & Vargas, L. (1991). *Prediction, projection and forecasting*. Boston: Kluwer.

Schwefel, D. (1987). *Indicators and trends in health and health care*. New York: Springer-Verlag.
Thirtle, C., & Ruttan, V. (1987). *The role of supply and demand in the generation and diffusion of technical change*. New York: Harwood.
Werbel, J. (Ed.) (1991). *Marketing: What's new, what's next*. New York: Conference Board.

Chapter 14

Competition

In any commercial enterprise, competition is inevitable. As soon as someone conceives of an idea for making money, others follow and try to capture their share of the potential market. This holds true in health promotion as well. Health promotion programs are subjected to competition from profit-oriented enterprises as well as from sources completely unrelated to health.

Health promotion managers and programs are often evaluated on the basis of the number of program participants. Because competition can reduce program participation, it is important to understand some of the forces that compete for those participants.

Continuing the open market analogy, we will consider competition from several standpoints. In the context of this chapter, we will define competition as any option that a potential program participant could choose instead of the health promotion program.

We will consider competition in two broad categories: positive and negative. Positive competition is any option that an individual could choose instead of participating in the health promotion program but that at the same time is congruent with the goals of the program. Negative competition is any option that an individual could choose instead of the program but that does not contribute to the goals of the program. Understanding these forces will allow managers to develop a competitive edge for their programs.

POSITIVE COMPETITION

Positive competition can be viewed in two ways. First, it can be viewed as a threat to the success of a health promotion program because it keeps employees from participating. Second, positive competition may be seen as complementing the services offered by the health promotion program. If employees can achieve their personal health goals outside the organization's program, the organization benefits without significant expenditure of its resources. This allows managers to allocate those resources to other areas of need. This second view of positive competition is the ultimate goal of health promotion: employees taking charge of their own health behaviors and making personal decisions to lead healthy lives.

Positive competition can come from commercial health and fitness clubs, community-based health education programs, and personal health-related activities.

Commercial Health and Fitness Clubs

Health clubs abound in most major metropolitan areas. Ranging from local YMCAs and small independently owned clubs to major franchises and chains, they can be a significant source of competition for a worksite health promotion program's fitness activities. However, although they may be able to help people meet most fitness-related goals, they are not able to address many needs for health education.

Health clubs compete with the organization's fitness center when they offer employees one or more of the following:

- Lower rates
- More convenient location, such as near one's home
- More convenient hours
- More or better equipment
- Affordable family memberships, especially if the health promotion program does not allow dependents to participate for free
- Attractive amenities such as massage, whirlpool, sauna, or tanning
- Daycare for children or other dependents
- Racquet sports
- Higher-quality instruction
- Better service, however that is defined by the consumer

If a local health club begins to draw participants from the program's fitness center, the manager should analyze the situation and ascertain whether it is possible to respond competitively. It is unlikely that the primary reasons for losing members will be fees or location. Here are some strategies that may help meet the challenge of the competition:

- Expand the hours to allow participants more flexibility.

- Provide increased personal attention both at the front end of the program and on an ongoing basis. This can involve more thorough orientation sessions, smaller groups, staff competitions to learn participants' names, and similar activities.

- Create special-interest programs to meet the needs of smaller segments of the population. Examples are marathon training, weight control, fitness activities for pregnant women, preseason conditioning programs for skiers, and programs for retirees.

- Offer activities involving participants' family members such as father-mother/son-daughter training programs.

- Integrate the fitness center with broader health education programs such as hypertension control.

- Integrate the fitness center into longer-range incentive programs. Placing higher-value incentives on the use of the worksite fitness center over community-based clubs may create a "hook" to retain members.

A health promotion program that does not have a fitness center may be able to establish cooperative programs. This can be accomplished by promoting membership in nearby clubs or arranging special corporate rates and company night activities. Before entering into any agreement, the manager should research the club thoroughly to be sure it is operating safely, professionally, and with a philosophy that is consistent with that of the health promotion program.

Community-Based Health Education Programs

A second source of positive competition is the wide range of health education offerings that are available in most communities. These include the following:

- YMCA and YWCA programs
- Education programs through the public school system
- College and university classes
- HMO programs
- Hospital-based community outreach efforts
- Commercial weight control programs
- Some health clubs
- Mental health providers

These facilities offer a wide range of programs that range from free to costly. They may appeal to the employee population for a number of reasons:

- Proximity to the home rather than work
- Availability of child care
- Independence from the workplace, which may appeal to the desire for privacy
- Highly visible and effective advertising and promotional campaigns

To counter these programs' competitive effect, health promotion managers can consider these changes:

- Scheduling classes in an "extended campus" format at convenient community locations rather than at the workplace
- Arranging for on-site child care
- Increasing internal promotional efforts using the same techniques found effective in media advertising
- Including participation in worksite programs in the organization's overall incentive strategy
- Arranging for community-based organizations to offer their programs at the worksite

Before expending much time and effort trying to counteract effects of community-based programs, managers should weigh their contribution to worksite program goals and their positive impact on resource allocation. Community health education efforts often are a welcome complement to worksite services.

Personal Health-Related Activities

In any organization there are people who do not participate in the health promotion program because they have a personal regimen that is meeting their needs. This group includes both low-risk people such as lifetime exercisers and high-risk people who are under a physician's care. Like other forms of positive competition, personal health pursuits generally support the same goals as worksite programs.

Personal Health Regimens

Employees who are pursuing healthy lifestyles are runners, walkers, swimmers, bikers, amateur nutrition experts, and meditators. They are at or near their ideal weight, know their cholesterol numbers, and exercise to videos in their homes. However, they seldom or never attend scheduled health promotion activities.

The members of this healthy population are the very role models the program would like all employees to emulate, but because they attend no program activities, they are invisible to the rest of the employee population. If they did participate, their high level of performance might intimidate less fit employees and discourage them from participating.

There are ways these health-conscious individuals can be integrated into the program:

- Design the overall incentive program to include credit for activities done outside the workplace.
- Host special events to attract these people, such as a 10K run and a 5K fun run followed by a race for all levels.
- Assist knowledgeable employees in designing and delivering brown bag seminars on topics within their purview.
- Recruit these employees to serve on advisory committees, planning councils, or as activity group leaders.
- Encourage them to contribute to an employee newsletter or serve as volunteers at health screenings.

It can be particularly inspiring to high-risk participants if employees who have successfully made significant lifestyle changes serve in leadership roles in the program. Their example can do more to motivate their peers than most incentive programs.

Participants Under a Physician's Care

A second group of employees who present positive competition are those who are pursuing healthy lifestyles as part of a medically supervised program. This group includes the following:

- The severely obese, for whom standard weight control classes are inappropriate
- People recovering from heart attacks who have a highly structured cardiac rehabilitation program
- Hypertensive employees for whom most activities are acceptable

These people may choose to participate in worksite programs, but they are likely to prefer more private settings. They may be motivated more by fear than by the desire to lose weight or improve fitness. They may not participate in health screening activities because they undergo complete medical examinations on a regular basis.

It is important to respect these employees' desire for privacy and anonymity. They must be brought into the worksite program on their own terms. They may want to pursue their health-related activities in private but use the program for the support it provides. They may participate only in specific seminars or general-interest educational activities. In any case, their level of participation should be given credit in the overall incentive program.

NEGATIVE COMPETITION

Negative competition is any factor that prevents an employee from participating in a health promotion program and does not contribute to the achievement of overall program or individual health goals. There are three kinds of negative competition: activities or circumstances that compete for an employee's time, activities that an employee inappropriately defines as fitness activities, and work-related activities.

Competition for Time

Although everyone has the same 24 hours in a day, people make very different choices about how to use their time. Some choices are truly voluntary; others are made out of necessity. Examples are single parents, people who need to hold two jobs, and people who commute in car pools. These populations have no less need for health promotion services than other employees and in many cases have more need. However, they are often more concerned about economically driven choices that compete for their time.

Second Jobs

There is little a worksite health promotion program can do for employees who must leave immediately after their day job to go to a second job. The primary strategy is to analyze their needs carefully and design activities they can do at home or at another convenient location, or during discretionary time such as lunch hours. The expectations of these employees and of health promotion managers must match their ability to participate. Credit should be given in an overall incentive plan for their modest levels of involvement. Shorter time commitments and alternative learning methods such as videos also can be useful.

Car Pools

Organizations that employ significant numbers of people from bedroom communities often have large numbers of car pools. They also are in areas where public transportation is minimal and traffic or parking is a problem.

To gain the participation of one car pool member, the manager must recruit each member of the group. Some of the strategies recommended for use with employees who hold second jobs also are useful with car pool participants. Employees who have extremely long commuting times on public transportation present similar challenges.

Non-Fitness Sports Activities

A second source of negative competition is recreational sports activities that appear to be fitness related but actually are of little benefit in this area. This includes such company-sponsored activities as bowling, softball, and volleyball leagues.

Many employees may believe that these activities are meeting their fitness needs. In fact, for people who have been quite sedentary, these activities may be a reasonable point of entry. In reality, however, many people participate in company-sponsored sports to meet social rather than fitness needs. This fact is the key to reaching these employees and bringing them into a health promotion program.

By building a significant social component into fitness classes, it is possible to attract many socially oriented employees. Group fitness classes that allow for interaction may prove attractive. Another approach is to combine a social event with a fitness activity that people usually do on their own. For example, runners who may not train together on a daily basis can schedule a weekly group run at a local park. Afterward they can share beverages around a picnic table or have a barbecue or pot luck meal.

Annual social events also can help meet employees' social needs. A moderately difficult bike ride at the end of the summer could wind up the biking season, especially in northern regions. The event can be conducted on a Saturday and may include families. Mini-triathlons, 10K runs, and racquetball tournaments also can be used as season-ending or monthly events with social aspects.

Work-Related Activities

The third source of negative competition is work-related activities: overtime, business travel, and company-sponsored events.

Overtime

Health promotion professionals must remember that the primary purpose of a worksite program is to benefit the sponsoring organization. When a choice must be made between the needs of the business and the needs of employees, in most instances the needs of the business will prevail. One example is that in periods of peak demand, employees may be required to work overtime.

Overtime affects management as well as hourly employees and unquestionably has an impact on health promotion programs. Employees are under additional stress both physically and mentally. The long hours affect them physically, and time away from their families and recreational pursuits affects them mentally. During overtime periods, employees must make choices about the use of their very limited discretionary time. At the end of a 10- to 12-hour shift, very few people will choose to stay longer at work to participate in a fitness class, even though it might be the best thing they could do.

For health promotion managers, the effects of overtime can be significant, particularly if they were not told in advance about the overtime decision. Programs will have to be rescheduled; even so, attendance will suffer. Overall participation will decrease. It may be necessary to discard promotional materials with dates and times, to renegotiate contracts with staff members, and to reduce hours for support employees.

Although the decision to schedule overtime is out of managers' control, there are some measures they can take to deal with the consequences.

- During the planning stages of a program, try to uncover work flow trends and anticipate peak periods when setting the program schedule.
- Make clear to the organization's decision makers that the health promotion department has a need to know about overtime scheduling and can be counted on to maintain confidentiality.
- Build into contracts or verbal agreements with employees a measure of flexibility to allow for circumstances such as overtime. If possible, spell out financial obligations.
- Plan alternative times and locations for classes.

Recognizing that during an overtime period the entire work force may be under severe pressure, health promotion managers can implement strategies that will meet the short-term needs of the population. These are some possibilities:

- Shorten the length of hour-long classes to a half hour.
- Offer 5-minute stress relievers during overtime hours, such as stretch breaks or partner shoulder massage.
- Offer additional brown bag seminars during lunch periods, noting that there may be a second lunch period for some shifts because of the extended hours.
- Increase the use of videotape checkouts for home education purposes.

The primary focus should be on meeting employees' needs, even if that means a lower level of programming during the overtime period. Additional high-pressure promotion is not appropriate and may hurt the image of the program.

Business Travel

A second source of work-related competition is business travel. This usually affects executives and sales representatives. Health promotion strategies to deal effectively with business travel center around alternative delivery systems. Some examples are the following:

• Offer more written educational materials.

• Offer audio tapes on appropriate topics.

• Investigate and promote fitness options that include reciprocity agreements among members of national organizations such as the YMCA, YWCA, and large commercial health club chains.

• Offer correspondence courses for education.

• Promote walking as a primary fitness option when traveling.

• Work with the organization's travel agency to recommend hotels that have fitness centers.

• Offer seminars to educate travelers about their unique problems:

  - Eating on the road
  - Combating jet lag
  - Fitness options in lieu of a "standard" workout

• Give credit for travelers' erratic participation in the overall incentive program.

Most employees who travel are salaried and well educated. As a group they tend to be receptive to health promotion concepts and welcome any extra efforts made to help them participate in the program.

Company-Sponsored Events

Company-sponsored social events may compete with health promotion programs. There is less a question of time than of the pressure or obligation employees may feel to attend such events. Senior management or department heads may send messages to employees that conflict with those sent by the health promotion program.

Here are some examples of competing company-sponsored events:

- Incentive programs that feature trips to resort areas; these often include open bars that encourage excessive alcohol consumption and traditional American meals with little consideration to nutritional goals
- Routine meetings where nonsmoking areas are nonexistent, inadequate, or not enforced
- Meetings that begin with continental breakfasts or breaks that offer participants poor nutritional choices
- Traditional office holiday parties, birthday celebrations, recognition events for promotions, or contract closings whose organizers fail to consider the mission of the health promotion program when planning refreshments

If the organization tends to conduct events similar to those described above, health promotion managers can make inroads into these traditions by meeting individually with the sponsors of the events and beginning an educational process to bring them in line with the goals of the program. Here are some possible approaches:

• Offer to plan menus for events to provide some healthier choices.

• Work with the catering departments of hotels that host events to provide healthier food options.

• Work with decision makers to move away from open bars in favor of a cash bar, which will encourage moderation in alcohol consumption.

• Work with decision makers to develop or expand smoking or clean air policies to include all company-sponsored events.

• Target the people who organize these events for personal attention and indoctrination into the health promotion program based on their personal needs.

• After successful indoctrination, include these people on health promotion planning teams or in the long-range planning process.

It is difficult to effectively administer a health promotion program and achieve long-range results when employees are receiving conflicting messages from management. Taking measures to ensure consistent support from management is an important step in creating a healthy worksite culture.

DEVELOPING A COMPETITIVE EDGE

Developing a competitive edge involves positioning a health promotion program in employees' minds so it compares favorably with the competition. For example, a vendor of weight control programs may guarantee quick weight loss. Health promotion managers can position their programs as permanent weight loss programs and support them with both incentives and a guarantee.

The first step in developing a competitive edge is to identify the program's competitors, what they offer that may have greater appeal to employees, and what real benefits employees will derive that the health promotion program does not offer. It may be useful to chart these factors on one or more grids as illustrated in Figure 14.1 on page 131. This visual representation will facilitate comparisons, and strengths and weaknesses will become more apparent. The second part of the process is to modify the existing program to meet or exceed the competition in the relevant areas. The final step is to promote the competitive edge consistently to employees.

Most health promotion programs are of high quality and effectively meet their population's needs. In today's marketplace, however, the competition for these consumers is constantly increasing. To maintain a program that receives the full support of both management and employees, managers must achieve and "sell" a competitive edge.

SUMMARY

Health promotion programs encounter many forms of positive and negative competition for their services. Positive competition comes from commercial health and fitness clubs, community-based health education programs, and individual health and fitness activities. Negative competition comes from second jobs, car pools, non-fitness-related sports activities, and work-related activities such as overtime, business travel, and company-sponsored events.

Managers can implement several strategies to gain a competitive advantage for the health promotion program. To be competitive, it is necessary to develop a marketing strategy. The first step in this process is to research and segment the market. We explain this process in the next chapter.

KEY TERMS

competition (p. 123)

SUGGESTED RESOURCES

Barabba, V., & Zaltman, G. (1991). *Hearing the voice of the market: Competitive advantage through creative use of market information*. Boston: Harvard Business School Press.

Day, G. (1984). *Strategic market planning: The pursuit of competitive advantage*. St. Paul: West.

COMPETITIVE ANALYSIS

Factors	*In-house Program*	*YMCA*	*Health Club A*	*Health Club B*
Proximity to worksite	0 miles	.5 miles	1.5 miles	1 mile
Price	$5/mo	$20/mo	$17.50/mo	$18/mo
Staff to participant ratio	1/200	1/450	1/300	1/300
Staff qualifications	High	Low	Low	Med
Hours	6am-6pm	6am-10pm	6am-12am	24 hours
Nautilus	Yes	Yes	Yes	Yes
Indoor track	No	Yes	No	Yes
Stairmasters	Yes	Yes	Yes	Yes
Bikes	Yes	Yes	Yes	Yes
Weight control	Yes	Yes	No	No
Stress management	Yes	No	No	No
Nutrition counseling	Yes	Yes	No	Yes
Smoking cessation	Yes	Yes	No	No
Body composition testing	Free	No	$5	$5
Hot tub	No	No	Yes	Yes
Sauna	No	Yes	Yes	Yes
Massage	No	No	Yes	Yes
Tanning	No	No	Yes	Yes
Child care	No	Free	$2/hr	$2.50/hr
Overall rating	2	1	4	3

Figure 14.1 Chart benefits on a competition analysis grid.

Hanan, M. (1991). *Tomorrow's competition: The next generation of growth strategies*. New York: American Management Association.

Kelly, J. (1987). *How to check out your competition*. New York: Wiley.

McAuliffe, R. (1987). *Advertising, competition and public policy*. Lexington, MA: Lexington Books.

Oster, S. (1990). *Modern competitive analysis*. New York: Oxford University.

Reis, A. (1986). *Market warfare*. New York: McGraw-Hill.

Schnaars, S. (1991). *Marketing strategy: A customer-driven approach*. New York: Maxwell Macmillan International.

Chapter 15

Market Segmentation and Research

Market segmentation is the process of dividing an entire market into subgroups based on characteristics that are relevant to the sale and distribution of a product or service. Segmentation is a useful technique because it is extremely difficult to design one service to meet the needs of all market segments. To determine the appropriate market segments for a health promotion program, it is necessary to conduct market research. This involves gathering information that will help managers understand the marketplace and its needs.

THE NEED FOR SEGMENTATION

The need for market segmentation in health promotion becomes apparent when certain activities are examined. Aerobic exercise classes are an excellent example. Aerobics classes can be designed at several intensity levels, with varying difficulty of choreography, for various durations, and with a wide variety of music. When a class is promoted without being targeted to a specific market segment, an extremely heterogeneous group of participants may show up. By the end of the class, in addition to some satisfied people, there undoubtedly will be participants who are exhausted, physically unchallenged, frustrated by the choreography, or bored with the monotony. As the class continues, those who are unhappy with any specific aspect will begin to drop out. At the end of the normal 6- to 8-week session, the small numbers of remaining class members will represent the market segment for which the class was best suited.

The damage to the program's reputation, the cancellation of classes because of drop-outs, and other repercussions all could have been prevented if managers had considered market segmentation before selecting and designing the class, and especially before promoting it.

INTRAORGANIZATIONAL SEGMENTATION

Intraorganizational segmentation refers to subdivisions within a single organization. The categories that are useful in health promotion are demographics, risk levels, and the relationships among these factors.

Demographic Segmentation

Demographics information is specific data about easily identified characteristics of a population. Demographic characteristics useful in planning a market-driven health promotion program are

- age,
- sex,
- job type,
- education level,
- income level,
- marital status, and
- family status.

The raw numbers of people in any category may indicate whether or not a specific activity is viable. A category's percentage of the total population may indicate whether there should be a major focus for that group. Demographic segmentation is critical in marketing activities to the target group. Materials, depth of content, and promotional and communication techniques all are influenced by the characteristics of these demographic segments. Choosing target segments, however, requires looking more closely at the needs of the organization in the areas of health risks and their impact on health claims.

Age

The number of potential participants in various age groups is one of the primary demographic factors in a population. Age is a consideration in relation to risks as well as interests. For example, older employees may feel more vulnerable to heart disease and may be highly motivated to learn about preventive measures. This is particularly true if a peer has recently suffered a heart-related problem. In contrast, younger employees may be more externally focused and may look to the health promotion program to meet some of their social needs.

If retirees are eligible for program participation and large numbers of retirees are located nearby, this can have a significant impact on program offerings. Retirees often want to maintain a connection with the workplace, and a health promotion program is often a vehicle that meets this need. Retirees' program interests are likely to be quite different from those of the working population. If their numbers are sufficient, health promotion professionals will need to identify and address their interests.

The greatest challenge is to attract large numbers of participants in several age groups. In this case it may be necessary to develop several program tracks throughout the year.

Sex

The numbers of males and females in a population are another demographic factor with important program implications. Managers will find significant differences in shared risk areas such as heart disease as well as health habits such as exercise and smoking. In addition, there are many sex-specific health issues, such as premenstrual syndrome (PMS), osteoporosis, menopause, prenatal health, and testicular cancer. These topics merit special emphasis if there are large numbers of either sex in the population.

There also are some generalizations about sex-based differences in attitudes and perceptions. For example, aerobic classes that are highly choreographed have a stronger appeal to women. Also, in most settings women tend to assume more child care responsibilities than men and as a result may have scheduling problems. This may drive decisions about time and location of activities or indicate the need for child care during activities.

Job Type

Job types are the broad categories of employment that exist in a work force. Examples are

- clerical,
- management,
- technical,
- sales,
- production, and
- administrative.

In certain industries there also may be specialized categories such as research and development.

Job types affect program decisions based on such factors as flexibility of work time, education level, amount of travel, and work cycles. Job type is related to education level and income level. Relationships among various population characteristics are discussed later in the chapter.

Education Level

The education levels of populations influence preferred modes of learning as well as the degree of receptivity to health promotion concepts. Better-educated employees tend to be open to a variety of educational modalities but welcome the use of written media. Less educated participants may prefer personal presentations or videos.

Income Level

Participants' income levels affect their ability to afford quality walking or running shoes or activity copayments. Their economic status also may determine whether they hold second jobs or have other factors that prevent them from attending program activities.

Marital Status

Marital status is a significant factor for several reasons. First, program participation will increase significantly if spouses are eligible. Second, it is wise to include spouses because of the profound impact they have on health care claims. Third, as they make changes in their lifestyles, employees may need to count on their home environment for support. Particularly in the areas of smoking cessation, nutrition, exercise, and weight control, married employees need assistance from their spouses in supporting the change. Health promotion managers can encourage spouse participation through home mailings and integrating community-based activities into the worksite program.

Family Status

Family status has an effect similar to that of marital status in that dependents represent a significant portion of overall health claims experience. For health promotion managers, the question of family participation presents two challenges. First, if children are to be included in the program, managers must consider an entirely different set of health education issues and learning modes. Second, if children are not included, managers may need to address the issue of child care while parents are attending program activities.

For planning purposes, it is critical to have current information on the numbers and percentages of an organization's population that fall into each of the demographic categories discussed above. Managers should gather these data early in the planning process and keep abreast of changes that may affect the program.

Relationships Among Demographic Factors

As mentioned earlier, program content is influenced by the relationships among specific characteristics of a population. This is particularly true of demographic data. For example, in planning programs for specific segments of a population, managers need to consider the relationship between age and gender. Women over 40 will have considerably different interests and motivations than men under 25.

Risk Factor Segmentation

Understanding participants' needs with respect to lifestyle change involves gathering and analyzing some very specific data. Segmentation by risk factor is a basic technique for designing program content. Decisions about what lifestyle classes to offer and how often depend on the specific risk factors found in the population and their prevalence, as well as the associated health care claims problems. Segmenting by risk factor is relatively easy and will result in a more effective program with higher utilization rates.

Health Risks

It is important to determine specifically, to the greatest extent possible, what numbers and percentages of the total population fall into each of several health risk groups. Some of the important risks are

- heart disease,
- stroke,
- obesity,
- smoking,
- stress,
- cholesterol,
- cancer,
- low back problems,
- nutrition,
- low birthweight and pregnancy-related problems,
- fitness, and
- women's health issues.

Health Care Claims

The second risk-related area involves not which risks the population has, but which ones cost the organization money. These data fall into many of the same areas as the personal health risks do, but may be described in terms of diagnostic categories rather than general lay terms. The data may become more complicated since often the cost of a risk area is camouflaged by the benefits design. One strategy that can be employed in a benefits plan is to decrease the coverage or increase the copayment for a problem area. This will have the net effect of lowering the cost to the organization while failing to get at the real problem. For this reason, it is important to look at risks from the perspective of both the individual and the organization in order to meet the needs of both customers.

Employee Interest Data

Thus far we have focused on participant needs, because the overall goal of marketing is to meet customer needs. However, when customers do not want what they need, the art of marketing comes into play to help them want what they need. For this reason, managers will find it useful to research employees' areas of interest.

Interests tend to be related to needs. Often, however, interest in an activity centers on the fun or social aspects of the activity rather than how it relates to a risk or need.

When planning a health promotion program, it may be appropriate to offer some activities that cater purely to the employees' interests without regard to needs. Group bowling or volleyball outings are examples. Such activities may be a point of entry for high-risk people who are threatened by the prospect of changing their lifestyle. By accepting these employees on their terms, managers can create a relationship with them and lay the groundwork for moving them gradually into programming appropriate for their needs.

Relationships Among Segments

After conducting market research, managers can organize the data into a grid (Figure 15.1) with any two categories of interest on each of the two axes. By constructing a grid, managers can see clearly the relationships among market segments and risk factors. For example, in Figure 15.1 it is clear that low seat belt usage is a major risk for the younger male population. It also can be seen that over half of the males in the work force fall into the two youngest age groups. The grid also shows the following:

- The largest group of employees is the male population (57.9%).
- The largest number of females is in the older age groups (18.4% of the entire population).
- As would be expected, stroke and heart disease are significant risks in the older male population. However, this is a relatively small group when compared to the rest of the population.

Other relationships it may be useful to examine are

- sex and age by education level,
- sex and age by job type, and
- sex and age by marital/family status.

The distribution of a population in any group may be the result of chance or of the relationships among demographic characteristics and risk factors. For example, it is probable that a relationship will be found between age and body weight, or sex and heart disease. Managers can use this information to identify market segments for programming purposes.

INTERORGANIZATIONAL SEGMENTATION

A second way to segment a population is by interorganizational criteria. Examples of these criteria are

- subsidiaries;
- regional centers of significant population;
- remote or small population centers; and
- organizational groupings such as divisions, product families, or functional groups.

Interorganizational units can be viewed as microcosms of the entire population, or they can be considered as completely separate and distinct populations.

Interorganizational Units as Microcosms

In populations that do not have vastly different operational functions, discrete field offices initially may appear to be candidates for segmentation. However, they may simply be smaller units of the population as a whole with similar demographics and risk factors. Examples of such organizations are

- field sales offices,
- U.S. Forest Service ranger stations,
- fast-food or similar retail chains,

MARKET SEGMENTATION GRID

Percent at Risk by Risk Factor (10-Year Projection)

Demographic Category	*Total Pop.*	*Percent of Pop.*	*Seat Belt Usage*	*Lung Cancer*	*Colon Cancer*	*Heart Disease*	*Stroke*	*Abnormal Pregnancy*
Male 18-25	75	17.7%	70%	8%	5%	10%	5%	0%
Female 18-25	21	5.0%	45%	10%	3%	5%	2%	28%
Male 26-32	81	19.1%	60%	12%	8%	15%	7%	0%
Female 26-32	30	7.1%	40%	15%	4%	8%	5%	32%
Male 33-40	57	13.5%	35%	20%	15%	25%	14%	0%
Female 33-40	49	11.6%	25%	22%	8%	17%	10%	45%
Male over 40	32	7.6%	25%	37%	19%	39%	43%	0%
Female over 40	78	18.4%	20%	30%	9%	24%	20%	30%
Total Male	245	57.9%						
Total Female	178	42.1%						
Total Population	423	100.0%						

Figure 15.1 A market segment grid helps define the specific market.

- a chain of banking institutions, and
- a large school district.

Using the school district as an example, virtually every school in the district will have a group of teachers, custodians, clerical and administrative staff, a school nurse, and food service workers. With the exception of the employee population and minor differences in the ratio of the sexes, most schools will look very much alike in most respects. For this reason, unless managers have other information, segmentation by location may be inappropriate.

There are exceptions within these examples. A fast-food chain may have several distribution warehouses and its own trucking operation. Of course, even in a largely homogeneous organization, the headquarters location probably will be unique. However, if there are many similar locations, the challenge facing health promotion managers may be related more to logistics and communication than to program decisions that reflect unique market segments.

Interorganizational Units as Distinct Segments

Large organizations may have multiple functions that must be treated as distinct, unique segments. For example, a large manufacturing organization may have subgroups such as

- a headquarters complex or campus,
- research and development laboratories,
- engineering departments,
- testing operations,
- subassembly plants,
- main assembly plants,
- customer service functions,
- continuation engineering groups, and
- a marketing and sales organization.

These units may be located near each other, or they may be scattered across the country or even around the world.

In such an organization, each functional group or physical location may need to be treated as a stand-alone organization with its own program plan. A complex organization like this is one of the most challenging assignments that can confront a health promotion manager.

Other examples of organizations that may require extensive segmentation are

- city, county, or state governments;
- large diverse holding companies; and
- community-based health promotion programs.

SOURCES OF DATA

There are a number of sources of data for use in health promotion market research. Some data can be found within the organization; other information is in the public domain.

Sources Within the Organization

Within the organization, managers can obtain data from the human resources department and by conducting research throughout the organization.

Sources Within Human Resources

Most human resources departments collect large amounts of data for various purposes: compliance with laws, defense in potential litigation, and planning.

In large organizations, most data are stored on computer databases and can be retrieved fairly easily. Many human resources reports contain the same information that health promotion managers need. Managers also may be able to obtain information from the employee assistance program or the insurer or third-party administrator (TPA) that pays health insurance claims.

These are examples of internal reports and their contents:

- Human resources database reports contain demographic information about the entire employee population and occasionally dependents.
- Aggregate EAP utilization reports contain information on the kinds of problems employees present to the EAP. These data can be useful in identifying stress management concerns.
- Insurance company or TPA reports contain detailed information about the health risks that cost the organization money. The relative importance of a risk is shown by the number of claims filed or total dollars spent. These data may need to be interpreted by an employee benefits expert.
- Occupational health and safety reports contain accident records. These are especially useful in developing low back injury prevention and stretch break programs.

In smaller organizations, most of these data will be available but may be found in less sophisticated formats, such as hard-copy files, that take longer to retrieve than computerized records. Once gathered, the information should be updated at least annually. To facilitate this process, managers may want to establish systems to simplify future retrieval efforts. This might be a joint effort of the human resources manager and the health promotion manager to develop a basic database.

Other Internal Sources

Additional research may be necessary when there are gaps in the data in standard reports. These data can be gathered through

- surveys,
- interviews,
- focus groups,
- on-site physical inspections, and
- library research.

Much of this additional research can be accomplished simultaneously in the overall data gathering process. For example, during a preliminary interview with the chief executive officer, the manager may obtain answers to specific questions and also be directed to existing reports or other resources in the organization.

In gathering information, interviews generally are most useful with individuals, focus groups with small groups, and surveys with large groups. Some of the people to be questioned are

- senior managers,
- human resources specialists,
- employee benefits managers,
- medical directors,
- occupational health specialists,
- safety managers,
- employee assistance coordinators,
- group insurance carriers, third-party administrators, or other claims payors,
- facilities manager,
- heads of significant departments that appear to have special needs,
- employee services or recreation managers,
- food service managers,
- random groups of employees, and
- representatives of special-interest groups such as runners or Alcoholics Anonymous.

Sources in the Public Domain

In some cases a manager is unable to obtain sufficient relevant data within the organization. An alternative is to use generic data from public sources and generalize to the organization from the populations described by those data. For example, if one knows that 30% of the general population smokes, it is reasonable to assume that the population of an organization that is demographically similar to the general population also will have approximately a 30% smoking rate. If there is a significant difference in demographics, other data may be compatible, or some assumptions may be made. For example, if the group is primarily male and over 50 years old, the 30% figure may be too low.

There are many sources for useful health statistics:

- Centers for Disease Control (CDC), Atlanta, Georgia
- National Institutes for Health (NIH), Rockville, Maryland
- National Center for Health Statistics, Beltsville, Maryland
- American Heart Association, Dallas, Texas
- American Cancer Society, Atlanta, Georgia
- Other health-related special-interest groups
- Federal government agencies (See Suggested Resources at the end of the chapter)
- State, county, and local health departments

SUMMARY

Using the process of market segmentation enables health promotion managers to offer services that are designed to meet participants' needs and to attract the targeted employees to those activities. The most useful method of market segmentation is based on demographic characteristics and health risk factors. Segmentation also can be done by geographic location, division, or other grouping.

In a health promotion setting, market research involves gathering information about the characteristics, needs, and interests of the eligible population. This information is usually available within an organization. It is also possible to gather generic data from public sources.

After conducting market research and dividing the population into segments, managers are ready to begin developing a marketing strategy. This is the subject of chapter 16.

KEY TERMS

demographics (p. 134)

intraorganizational segmentation (p. 134)

job type (p. 135)

market segmentation (p. 133)

SUGGESTED RESOURCES

Breen, G., & Blankenship, A. (1989). *Do-it-yourself marketing research.* New York: McGraw-Hill.

Dickinson, J. (1990). *The bibliography of marketing research methods.* Lexington, MA: Lexington Books.

Green, P., Tull, D., & Albaum, G. (1988). *Research for marketing decisions.* Englewood Cliffs, NJ: Prentice-Hall.

Michman, R. (1991). *Lifestyle market segmentation.* New York: Praeger.

Tull, D., & Hawkins, D. (1990). *Marketing research: Measurement and method.* New York: Macmillan.

U.S. Government Health Data Sources:

Department of Commerce, Bureau of Census	202/763-4040
National Technical Information Service	703/487-4600
Congressional Budget Office	202/226-2650
House Energy and Commerce Subcommittee on Health and the Environment	202/225-0130
Ways and Means Subcommittee on Health	202/225-7785
Senate Finance Subcommittee on Health	202/224-4515
Environmental Protection Agency	202/382-4454
General Accounting Office Human Resource Division	202/275-5470

Department of Health and Human Services:

Administration on Aging	202/245-0641
Alcohol, Drug Abuse, and Mental Health	301/443-4408
Food and Drug Administration	202/443-3170
National Center for Health Statistics	301/436-8500
National Institutes of Health	301/496-4000
Office of Disease Prevention and Health Promotion	202/245-7611

Chapter 16

Market Strategy and Product Planning

Every product in the open market goes through distinct phases of development that are known as the product life cycle. Managers apply marketing principles in planning for the changes that inevitably occur as the product life cycle progresses. By applying these principles, traditionally known as the four Ps of marketing, managers can continue producing products that meet the needs of their end users. This is a process that is closely tied to long-range planning. Health promotion managers can apply the same strategic planning techniques to their programs and services.

PRODUCT LIFE CYCLES

All products go through four predictable life cycle phases as shown in Figure 16.1: introduction, growth, maturity, and obsolescence. The characteristics of these phases give managers direction for health promotion planning activities.

Product Introduction

The first phase of the product life cycle is introduction. In manufacturing, this phase encompasses design and development activities, pilot testing, market testing, and a

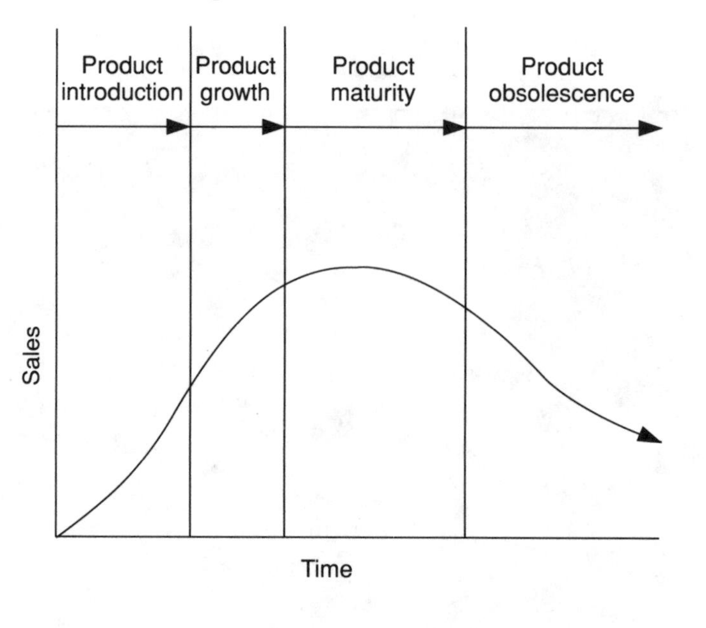

Figure 16.1 All products and services go through the product life cycle.

refinement process. During this phase, it is unlikely that the product will be profitable because of the resources being allocated to development activities and the relatively low volume of distribution.

In a health promotion program, examples of product introduction are:

- The entire period when a health promotion program is being introduced to a population
- The early stages of a new fitness center
- The organizing and planning phases of any new education program that is being designed to meet the needs of a specific population
- A change to a health risk appraisal product
- Introduction of a new vendor in any program area

During this phase managers can expect lower than normal participation due to inertia, and a resulting reduction in revenue if a fee is charged for the service. Later in the chapter we will examine strategies to counteract these effects.

Product Growth

The second phase of the product life cycle is growth. This is when advertising and promotion efforts are strongest. Distribution channels are in place, and sales volume is building. Competition has not yet become a factor, and maximum prices can be charged. As a result, per-unit profit is at a peak.

In health promotion programming, evidence of the growth stage of an activity includes:

- High levels of excitement about that aspect of the program
- High visibility in newsletters, bulletin boards, flyers, and other internal media
- Little evidence of participants moving to alternative competitive activities

- Little hesitation to pay the fees charged for the activity
- Increasing participation rates with no evidence of leveling off

This is a time when managers may believe they cannot lose with this activity. It appears to be ideally matched to the needs of the population.

Product Maturity

The third stage of the product life cycle is maturity. The first evidence that this stage is approaching is the onset of competition. Competing service providers have caught on to this opportunity and have begun to introduce their own versions of the product. The product is being fine tuned, and the effects of the market, such as supply, demand, and competition, are being evaluated. Sales volume will peak during this phase until competition begins to erode market share.

In a health promotion environment, evidence of maturity includes these events:

- Hearing about participants who have attended similar activities in the community, often with a slight creative difference or a lower price
- Seeing an end to the rise in participation rates
- Beginning discussions about how to spice up the activity to give it a new look

As a program activity matures, managers must begin planning to introduce the next generation of the activity. This process must begin in a timely manner so that there is little or no time between the final phase of the first activity and the introductory phase of the next.

Product Obsolescence

The final phase in the life of a product is obsolescence. This phase is characterized by declining sales volume and a saturated market. As volume declines, the cost per unit tends to rise, which erodes profit margins. This causes managers to focus on cost efficiencies to help maintain margins. The marketing department is challenged to begin product innovations that will keep volume up as long as possible. This is often when "new and improved" products are promoted. Strategic planners are already planning new products to replace the original when the timing is right.

In health promotion programs, obsolescence is evidenced by various signs:

- Classes being run with fewer than an optimal number of registrants
- Cancellations because of low registration
- Cost per enrolled participant rising
- Discussions centered around the "market being saturated" while empirically it is clear that there is plenty of opportunity, for example weight control classes are not going well while overweight participants are very prevalent

Like manufacturers, health promotion managers should be planning both innovations to existing products and completely new future offerings. Understanding and planning for the four phases of the product life cycle will enable managers to maintain maximum program participation. Responding creatively to the changing phases can bring new life to tired activities and enliven the entire program.

Factors Influencing the Length of the Cycle

Although the phases of the product life cycle are predictable, the lengths of the cycles can vary widely. Some of the causes of the variation are beyond the manager's control. In other cases, managers can use marketing techniques to influence a cycle's length and prolong the life of a product. Some factors that may be beyond a manager's control are:

- The presence of competition
- The potential for competition
- Turnover in the eligible population
- External influences such as the increased public awareness of cholesterol, which boosted interest in nutrition activities
- A highly proprietary product that cannot be readily duplicated, such as an innovative computer program for health risk appraisal

With an awareness of product life cycles and factors that can affect them, managers have an opportunity to control them to some extent through the use of marketing principles. In chapter 11 we presented six Ps of marketing in relation to health promotion. In the area of market strategy, we focus on the traditional four Ps of marketing.

THE TRADITIONAL FOUR Ps OF MARKETING

The basic principles of marketing are often characterized by the four Ps: product, price, promotion, and position. These categories reflect basic principles that allow managers to plan for changes in their services—both proactively in anticipation of the inevitable life cycle of the product and reactively to "save" struggling program activities. The four Ps often overlap and interact; it is frequently difficult to make adjustments in one area without affecting the others. Some of these interdependencies will become apparent in the following discussion.

Product

A core product that serves as the bread and butter of a business can often be rejuvenated to extend the life cycle or even recycled in the apparent form of a new product. The laundry soap industry, for example, has taken a staple of American households and continually recreated it in new and improved versions over the years, allowing large companies to compete for market share successfully while continuing to offer their core product. The switch from powdered detergents to liquid was a change in the basic product. This change effectively repositioned the liquid as more convenient. But the packaging of the liquid in plastic bottles spawned criticism from environmentalists. This then caused a change back to powder—in a new recyclable cardboard package that repositioned the product as environmentally correct. A change in the product was made to affect the position. An unanticipated market reaction forced a return to the original product in order to regain a favorable position. This example illustrates the interdependence of the four Ps. The addition of bleach crystals, more concentrated formulas, built-in fabric softeners, and various scents are other changes that have been made in the laundry soap product to meet the needs or demands of the market or to differentiate from the competition.

In a health promotion program, changes to a core product can be illustrated by a fitness class. A basic fitness class could be altered to meet the needs of an older population by

- emphasizing walking as a primary activity,
- adjusting lecture content to emphasize specific nutrition needs, and
- discussing the different social aspects of group fitness activities that are better suited for older people.

Likewise, the basic program could be repackaged to reach other specific markets:

- A "new and improved" body composition testing method could be used to regenerate interest in the activity during promotional events.

- The classes could be restructured from 1 hour a week for 6 weeks to two 3-hour seminars; this repackaging would also reposition it as a "seminar" rather than a class.
- A nutrition class could be changed from a generic version to one that emphasizes the nutritional needs of pregnant women.

Virtually any health promotion activity can be restructured, modified in content, or repackaged to breathe new life into the product.

Price

Price considerations take into account the market's willingness to pay for products. Certain market segments may be willing to pay more for similar products or services. For example, shoppers can purchase the identical can of soup or sack of flour from a "warehouse" supermarket at significantly lower prices than at a "full service" supermarket with wide aisles, large deli departments, and numerous free sample displays. Shopping at the more expensive supermarket is often perceived as a more enjoyable or convenient experience. The warehouse supermarket is viewed as one for the thrifty and wise consumer.

In health promotion programs, price can take on different meanings. Price may be the fee or copayment for a specific activity. However, in many programs which are subsidized, the price that a participant pays can take the form of their time, convenience, or even something more personal such as embarrassment.

Participant Fees

In the case of an actual cash payment, the price can be changed as a result of an early registration discount promotion. It can be put on an easy payment plan through the use of payroll deductions, or a fee for a formerly free program can be instituted to give the program a higher perceived value. Often products and especially services that are free are perceived as having very little value. The typical consumer when comparing two items that are difficult to differentiate such as tennis balls or water bottles, will sometimes choose the higher priced item with the expectation that "it must be better since it costs a little more."

Participant Time

In the case of a participant's time, classes that are too long may be too high a "price" to pay. For example, if an employee has a standard one-half hour lunch period, the hassle of getting permission from their supervisor and covering their work station to arrange for a one hour time block for a noon class may "cost" them too much. It may be possible to modify the content and repackage the class into 30-minute segments in order to meet the "price" requirements of the market.

Participant Convenience

Convenience can be a simple function of scheduling. If the organization has a large commuter population, staying after work and making car pool arrangements or missing public transportation schedules may be inconvenient. Populations that travel frequently may not be able to commit to classes that meet weekly for more than a few weeks. Activities that are scheduled at a significant distance from participants' work stations may require too much time traveling for the class to be held over the lunch period. Inconvenience may drive participants to the competition. If the fitness center is not open early or late enough, it may be more convenient for a participant to stop at a local health club. The "price" is lower.

Participant Embarrassment

The price of being embarrassed in the presence of coworkers can be a devastatingly high price for a participant to pay for an activity. An example of this situation is that of a fitness class targeted for the obese. If obese employees must show up at the class and risk being seen either wearing exercise clothing that they feel is unflattering or doing movements that make them uncomfortable in front of others, the "price" is too high. If this situation exists because the class is held in the cafeteria after work, perhaps the price can be lowered by providing a more private setting for the class.

Promotion

Promotion concerns when, where, and how information about a product or service is brought to the consumer. It is discussed in detail in chapter 17. Promotional vehicles include

- advertising,
- news releases,
- health fairs,
- direct mail campaigns,
- free samples, and
- special offers.

Advertising

Advertising typically is done via print (trade journals, magazines, newspapers, or direct mail) or broadcast media (radio and television). Also used are audio- and videotapes and computer diskettes with animated graphics.

In health promotion, advertising can be done via

- newsletter articles,
- posters,
- electronic mail, and
- table tents.

News Releases

A news release is a brief written statement that is sent to appropriate media to announce a new product or service or a change in an organization in a way that will draw favorable attention. News releases can be an effective way to receive free publicity. However, there is no guarantee that a news release will be printed, and news releases are rarely printed in full.

In health promotion, announcements of new employees, classes, and activities can be submitted for publication in the internal newsletter. It is reasonable to expect such items to be published.

Trade Shows

Trade shows are events that generally are held in conjunction with conferences for specific professional organizations. For example, at a conference for members of a state teachers association, the exhibits at the trade show would include vendors of textbooks, educational materials, school furniture, physical education equipment, and other products or services that may be of interest to teachers. The vendors pay a "booth space" rental fee and provide their own displays. The exhibit floor is usually available during the entire conference or may be closed during keynote speakers or other special conference events.

The health promotion equivalent of a trade show is the health fair. Health fairs can be organized around specific risk areas such as fitness, seasonal interests like summer sports, or be general in nature with information on a number of topics. Health fairs are fairly labor intensive to organize and may cause some disruption at the workplace, so they should be carefully scheduled and planned.

The cost of operating a health fair may be reduced through the use of vendors and public agencies as well as volunteers. Vendors often will staff health fair booths in exchange for their exposure to employees. The role of vendors must be clearly defined and may need to be approved by management. Local television or radio stations sometimes run health fairs for a modest fee to satisfy the public interest requirement of their FCC license.

Direct Mail Campaigns

Direct mail is one of the fastest-growing methods of promoting products and services. Correctly done it can be a highly effective means of reaching a target market. In the business world, direct mail is used as a technique to generate leads: people or organizations that express interest in a product or service.

In planning a direct mail campaign, the first step is to identify a target market by using criteria such as geography, demographics, previous buying habits, or topical interest.

The second step is to develop or purchase a mailing list for the target market. Lists can be obtained directly from trade groups, purchased through brokers, or developed through prize drawings at trade shows. For worksite health promotion programs, managers usually have access to employee lists, although there may be restrictions on home mailings. Managers can develop mailing lists from class rosters by topical area, drawings at health fairs, or interest surveys.

The third step is to develop the concept, including the attention getter, the message, the call for action, and the "spiff." The attention getter is the gimmick that tantalizes the recipient into opening the mail. An example is the overused "You may already have won!" message on magazine sweepstakes mailings. The message explains the features and benefits of the product or service and leads the reader to the call for action.

The call for action is the request for the reader to return a reply card or call a phone number to obtain more information and receive the spiff: a free incentive for following through with the call for action. The spiff can be an actual item such as a keychain or water bottle, or it may be an incentive, such as doubling the odds in a drawing for a grand prize. The reply card should have space for the recipient's name, address, and some qualifying information that can be used to identify valid candidates for the product or service.

In a health promotion direct mail campaign, qualifying information can include the following factors:

- Current percent of body fat or amount overweight to qualify for weight control programs
- Number of times per week one exercises for fitness classes
- Score on a simple 10-point stress quiz for stress management activities
- How many times one has tried to quit smoking for a smoking cessation program
- Whether one owns a bicycle

After the mailing has been sent, the fulfillment phase of the project begins. This involves recording all responses and fulfilling the promise to provide more information as well as the spiff. Based on the qualifying information, the respondents are separated into unqualified and qualified leads. The qualified leads are contacted to qualify them further. At this point they either become prospects or remain in a file for future promotional efforts. A prospect is pursued by a sales representative until a sale is closed or the prospect unequivocally says "No!" and is considered "dead." In health promotion, specific questions can be scripted for the follow-up contact to extract the desired information. Prospects are closed when they officially register for the activity being promoted.

Free Samples

Free samples are a highly effective method

of promoting a product or service. Product samples are mailed or distributed in person. In the service area, a free sample might be an introductory seminar on how to make money investing in residential real estate. The audience is given just enough free information to pique their interest in signing up for a complete course.

In health promotion, a free sample could be a lunchtime brown bag seminar where one portion of a course is given by the instructor who teaches the complete course. For example, a 20-minute seminar on "How to Determine the Fat Content of Foods Through Label Reading" can be a sample of a cholesterol reduction course. Other examples of free samples are

- a free body composition measurement,
- free low-fat recipes, and
- a free fitness assessment.

Special Offers

Special offers give consumers the opportunity to buy a product or service at a lower price and/or for a specified time period. Customers may be offered a free gift as an incentive to make their purchase within a specific period. Other specials are

- volume discounts,
- end-of-season sales,
- inventory reduction, and
- interest-free delayed-payment plans.

In health promotion, similar strategies can be applied. Examples are

- early-bird reductions in course fees,
- free water bottles for the first 30 entries in a fun run,
- two for the price of one education course incentives, and
- limited-time offers for computer nutrition analysis when the software is borrowed or rented.

Position

Positioning is the emphasis that is used to offer a product or service. Repositioning is offering the same product or service with a different emphasis. An example is a popular antacid that originally was positioned as an aid for indigestion and later was offered as a calcium supplement at a time when osteoporosis was being widely publicized. Another example is a shampoo that originally was positioned for use with babies because it did not irritate the eyes. This product was later repositioned as a woman's shampoo because of its gentleness.

In health promotion, there are abundant opportunities to reposition services. This technique can open the doors to new markets and extend the life cycles of various aspects of the program. Examples of repositioning are listed below:

- Fitness classes for runners instead of an older population, or as an adjunct to weight control
- Smoking cessation classes being offered as smoking reduction classes for those not quite ready to quit
- Weight control classes for prevention of cardiovascular disease as opposed to self-esteem or vanity
- Nutrition classes as part of prenatal care rather than cholesterol reduction
- Stress management as a personal productivity technique rather than a mental health tool

When a program is repositioned, it may be necessary to change the content and the methods of promotion. For example, the content of the nutrition class for cholesterol reduction would have to be modified to make it appropriate for pregnant women, and promotion techniques would have to be developed for that market.

SUMMARY

In designing health promotion activities, an understanding of the four phases of the product life cycle will allow managers to apply marketing principles in planning for the inevitable changes.

The four Ps of marketing are product, price, promotion, and position. When managers incorporate these concepts into the long-range planning process, they will find their programs enjoying stable participation with activities that remain viable for longer periods of time.

Because one of the four Ps of marketing, promotion, is included in the term health promotion, further discussion is warranted. The next chapter addresses this important area.

KEY TERMS

leads (p. 149)
position (p. 150)
price (p. 147)
product (p. 146)
promotion (p. 148)
prospects (p. 149)
qualifying information (p. 149)

SUGGESTED RESOURCES

Day, G. (1990). *Market driven strategy: Processes for creating value*. New York: Free Press.

Glaros, T. (1988). Applying marketing principles to worksite health promotion. *Fitness in Business*, **2**, 137-139.

Hillestad, S. (1991). *Health care marketing plans: From strategy to action*. Gaithersburg, MD: Aspen.

Jain, S. (1985). *Marketing planning and strategy*. Cincinnati: South-Western.

Thorelli, H., & Cavusgil, S. (Eds.) (1990). *International marketing strategy*. Oxford, NY: Pergamon Press.

Chapter 17

Program Promotion

In any health promotion program, one of the greatest continuing challenges is successfully promoting activities. Not only do managers need to provide solid program offerings with attractive themes and incentives, but also there must be a constant effort to keep participants informed about all aspects of the program. In this chapter we discuss program promotion in the context of the communication process.

Program promotion begins with an analysis of the organization's existing communication systems. Communication with management is an important element of program promotion. In many organizations, managers also must consider the public relations implications of their program communications. This is particularly true in smaller communities where an organization is a primary employer.

COMMUNICATION ANALYSIS

The first step in planning a promotional program is to analyze the organization's formal and informal communication systems. Other significant factors are learning styles, reading levels, job-related considerations, and the special needs of dependents.

Finally, managers must consider the organization's communication policies.

Communication Systems

In every organization there are formal and informal communication systems and special communication events.

Formal Systems

Every organization has formal systems of communication, as discussed in chapter 9.

These systems are the primary vehicle for promoting a worksite health promotion program. Some of the common formal communication systems are

- newsletters,
- bulletin boards or other posting areas,
- memos,
- meetings,
- flyers,
- paycheck stuffers,
- public address systems, and
- electronic mail.

These vehicles are discussed in more detail later in the chapter.

Informal Systems

Although an organization's formal communication systems are the official channels for a health promotion program's communication plan, informal systems often are more effective. Examples of informal systems are

- networks (already existing or specifically developed by health promotion professionals);
- rumors;
- word of mouth; and
- incentive-related communications such as T-shirts with a program message.

A great deal of information about the health promotion program can be disseminated through these informal channels. For example, participants who completed a successful "1,000 Mile Club" running event and were rewarded with a T-shirt will be seen wearing the shirts by others. The shirts can become a conversation piece from which the participants share their enthusiasm for the program in general or the event specifically. Word of mouth communications are extremely important. However, a participant's minor negative experience may be magnified to the detriment of the program. Managers should stay in touch with "the word on the street" about the program and take "damage control" measures when necessary.

Learning Styles

In communicating information about a health promotion program, it is important to consider differences in participants' learning styles. Within any population there are people who span the continuum of learning styles. Most styles are a function of preference or experience. By understanding the various learning styles, managers can communicate health promotion program information in ways that will reach the largest numbers of participants.

Information can be presented in a number of ways, including

- standard written format for participants with preferences for reading and detail,
- abbreviated written executive summaries for participants who prefer reading but have limited time or need for detail,
- in-person group presentations for individuals who prefer oral communication without significant interaction,
- in-person individual presentations for individuals requiring more explanation and interaction,
- videotaped material for participants who are visually-oriented,
- audiotaped material for people who prefer the spoken word but need flexibility in scheduling, and
- computer-based communications for those who need self-paced activities.

In addition, supplemental delivery systems may be required for populations with special needs such as

- signing for the hearing impaired,
- Braille translation or large-print media for the visually impaired,

- multilingual translation for populations where English is a second language, and
- special keyboards for computer-based education for people who are physically disabled.

Reading Levels

An important consideration in communicating program information is reading levels in the target population. If the target population is known to have low reading levels, it may be necessary to rewrite materials to meet their needs.

Job-Related Considerations

In communicating information about a health promotion program, managers must consider employees' work schedules and commuting and travel schedules. Considerations related to work schedules include

- hours/shift worked,
- length of work day,
- length of lunch and break times,
- need to work on an assembly line,
- inflexible deadlines,
- management policies, and
- work rules dictated by union contracts and/or job categories.

Considerations related to commuting and travel include

- travel schedules,
- primary work location, such as remote facilities or home-based work, and
- commuting distances.

To accommodate employees who may have problems in these areas, managers may need to offer several flexible program options:

- Classes organized into convenient time units such as consolidated seminar formats rather than extended weekly classes
- Modified class lengths, such as 1/2 hour rather than hour-long formats
- Increased use of self-study materials
- Use of video as a substitute for written media
- Informal networks and word of mouth communications
- Holding classes in locations closer to a group's work station to reduce travel time
- Communications directed at preexisting and underserved social groups rather than those of a general nature

Dependents

Dependents are often an overlooked group in worksite health promotion programs, in part because they are difficult to reach in the communication effort. Communications intended for employees may not reach dependents. There may be budget or policy restrictions that prevent thorough penetration of the dependent population. Suggestions include the following:

- Mailing information directly to the dependent at home
- Including program information in other communications, such as benefits information, that have a high probability of reaching dependents
- Offering promotional information whenever dependents are at the workplace, such as at open house events

Reaching dependents is a significant challenge because for most organizations they are major consumers of health care dollars.

Policy Issues

Many organizations have policies that influence the options for communicating program information to participants. These may include guidelines and approval procedures for both internal and external communications.

COMMUNICATION OPTIONS

The communication options available to health promotion managers are both traditional and nontraditional.

Traditional Communication Options

Some traditional communication options are listed on page 154. These methods, while well known, merit some discussion.

Newsletters

Newsletters are a common communication medium in organizations. Some are developed internally; others are purchased from vendors. Health promotion may choose one of these options or may choose to communicate program information in another way. Table 17.1 lists some advantages and disadvantages of three newsletter approaches.

There are several ways to use newsletters to communicate information about health promotion programs. The simplest and least expensive way is to prepare a health promotion column to appear in every issue of an existing in-house newsletter. The editor will probably specify an acceptable format and length and set a deadline for submission of copy.

After agreeing to provide a regular feature, it is important to keep the commitment. A good way to do this is to look at the plan for the year and write most of the feature articles in advance so that meeting publication deadlines is not a problem.

Providing a regular column to an existing newsletter is a good way to begin effective newsletter communication. It is often the method of choice for new programs or for programs in small organizations.

The next option is an in-house health promotion publication that is written, printed, and distributed by the health promotion staff with contributions from participants. This can be an exciting project; however, before making the commitment to produce a newsletter, managers should consider

- availability of qualified employees,
- frequency of publication in relation to available resources, and
- cost and quality.

After selecting this option, the manager should form a newsletter task force or advisory committee to take responsibility for the project. Some of the contributions that a participant-based group can make are

- overall planning,
- writing talent,
- editing talent,
- desktop publishing,
- photography and illustration, and
- printing and reproduction.

Two important decisions are the length of the newsletter and the frequency of publication. These factors will affect the resources needed to publish the newsletter on a regular basis.

A dedicated in-house health promotion newsletter can be a highly effective communication vehicle. It is often the method of choice for very large organizations with well-staffed health promotion departments.

The next option is to purchase a professionally prepared health promotion newsletter from a vendor. There are many

Table 17.1
Newsletter Options: Advantages and Disadvantages

Type	Advantages	Disadvantages
General in-house	No production expense Last-minute changes possible	Poor readership possible Poor placement of article possible
Dedicated in-house	Editorial control Employee commitment to program Low cost possible Inclusion of local program information	Low-quality product possible Consumption of nonmonetary resources Production deadlines
Professionally prepared	High quality Few administrative responsibilities Predictable deadlines Wide range of options	High cost Lack of editorial control Untimely content possible

newsletters available with a wide variety of options, including

- custom mastheads that include the organization's logo and program name;
- bulk shipping to the health promotion office or direct mailing to participants' homes;
- a variety of editorial approaches, ranging from a strictly scientific or medical point of view to lay-oriented perspectives;
- a variety of emphases, including fitness, nutrition, family, and balanced approaches;
- a full range of lengths, types of paper, color choices, and other production options;
- the ability to include local information on one or more pages; and
- the ability to include a local insert stuffed into the newsletter before distribution.

Bulletin Boards

Bulletin boards can be an effective communications vehicle or a complete waste of time and money, depending on how they are located, developed, and maintained. In many organizations the location of bulletin boards is regulated by the human resources or communication department. The health promotion department may be allotted some dedicated space on existing bulletin boards, be given a dedicated board, or have a wide range of posting options, including posting on nonbulletin board locations. Good locations for postings are outlined in Table 17.2.

Maintenance of bulletin boards is an important factor in their effectiveness (see Table 17.3). If bulletin board items are not changed regularly, people lose interest and stop thinking of the board as a source of useful information. Bulletin boards are a core communication method for health promotion programs and should be managed carefully.

Bulletin board materials can be produced by the health promotion staff for specific program needs, or they can be obtained from vendors. There are vendors that produce bulletin boards. Vendors of services at the workplace may provide posters specific

2
Board Location

Location	Examples
High-traffic areas	Time clocks Entrances Exits Lavatories
Idle-time areas	Copy machines Cafeteria checkout lines Water cooler/coffee machines
High-readership bulletin boards	Internal job posting boards Union announcement boards Recreation announcement boards

Table 17.3
Bulletin Board Maintenance

Quality

Use colors, not black and white
Use background material, not cork board or other permanent material
Use nonstandard size and shaped paper
Use attention getters like three-dimensional components
Use short, clear messages
Use artwork when possible

Maintenance

Use a specific take-down date for all postings
Replace frayed or torn materials
Regularly replace all postings
Regularly resupply "take one" materials
Regularly monitor readership to determine effective life of postings

to their services. Public agencies often have low-cost, attractive materials. For example, the American Cancer Society's Great American Smokeout provides high-quality, professionally prepared materials.

Group Communication Meetings

Group communication meetings can be a useful vehicle for health promotion communication. However, it may not be practical or possible to take employees away from work to attend a special health promotion meeting. It may be preferable to request a place on the agenda of a general meeting. This will give the health promotion program the same status as other items on the agenda and will confirm the department's established role in the organization.

Flyers

A flyer typically is a one-page announcement of program events. Flyers are often produced on colored paper with graphics and a fairly brief message. The information provided usually is in the classic "Who? What? When? Where? and How?" format. Having a convenient tear-off registration form can increase flyers' utility in the promotion process. An example of an effective flyer is shown in Figure 17.1.

The health promotion department can use a desktop publishing or word processing program to produce customized flyers, or they can be prepared by vendors.

Paycheck Stuffers

Paycheck stuffers are an inexpensive way to ensure that a small promotional piece will actually get in the hands of all employees. The use of paycheck stuffers usually must be coordinated with the human resources department, the payroll department, or both.

This communication method will lose effectiveness if paycheck stuffing is so common that participants perceive it as

Your Best Is Good Enough

Caregiving Series

Being a caregiver to the elderly seems to "just happen," even though growing older is a gradual and predictable process. The demands of caregiving catch many of us by surprise—like being thrown into a job you've never been trained for when you already have a job:

- *20% to 30% of employees over 30 years old have caregiving responsibilities.*
- *3 out of 4 caregivers are women.*
- *Caregivers spend from 12 to 35 hours a week on routines such as shopping, cooking, and finding care for the elderly.*

If you are involved with caregiving or stressed by fears for parents living at some distance, plan to attend this special 4-week series that will address the frustration and sadness caregivers experience, identify inner resources and external support services, and provide an opportunity to share and discuss coping strategies:

Week 1: Stress & Family Crisis—Tuesday, March 9

Week 2: Difficult Parents—Tuesday, March 16

Week 3: Family Dynamics—Tuesday, March 23

Week 4: Support Systems—Tuesday, March 30

All sessions will be held from noon to 1:00 p.m. in the Anniversary Room.

★ ★

Registration

Name ______________________ *Phone* ______________

Facility code ______________ *Business group* ______________

Sessions I plan to attend:

_____ *Week 1: Stress & Family Crisis* _____ *Weeks 3: Family Dynamics*

_____ *Week 2: Difficult Parents* _____ *Week 4: Support Systems*

Return to Health and Home Life Resources • HQS02G

Figure 17.1 A flyer can be an efficient way of promoting an event.

another form of junk mail. An alternative to paycheck stuffers is a printed message that appears on paycheck stubs. This technique may be appropriate for brief announcements.

Public Address Systems

Public address systems are not always available or appropriate for use in a health promotion program. They are most commonly found in large manufacturing settings, and there may be policies that restrict their use. A public address announcement is most useful as a last-minute reminder of an event that is about to occur or as an announcement given at the beginning of the day. Public address announcements can be used to tell employees about lunchtime events or major events such as health screenings.

Nontraditional Communication Options

In addition to traditional communication methods, there are several less traditional options, including portable resource centers, lavatory postings, answering machines and voice mail, and marquee announcements.

Portable Resource Centers

Placing materials in any of several portable settings can be a useful communication method. In large buildings, managers develop a resource center on a mobile cart with educational information, handouts, and registration materials. The cart can be placed in an elevator or moved to any strategic location that affords access to participants. The cart can be moved from one location to another during the day. It is important that the participants know in advance where the cart will be, and that the cart actually arrive on schedule.

Lavatory Posting

Posting promotional materials in lavatories is an effective method of communicating information about health promotion programs. In most cases management approval is required.

General information can be posted on the inside doors of lavatory stalls. Lavatory posting also can be used to promote activities that are specific to the sexes, such as breast self-examination or testicular self-examination. If there are scales in lavatories, information about weight control programs can be posted nearby.

A major drawback of lavatory posting is that it may attract vandalism and graffiti. This will require diligent maintenance on the part of staff members.

Answering Machines and Voice Mail

The use of answering machines and voice mail has become a standard practice in business settings. They are cost-effective and can be a useful communication vehicle. In a health promotion setting, a brief announcement can precede the instruction to leave a message. In more sophisticated voice mail systems, there may be some menu options that allow the caller to listen to a calendar of events, request more information about specific programs, or ask to be put on a mailing list.

Marquee Announcements

Marquees are not commonly seen in workplaces. However, if they exist, or if a special programmable marquee is purchased specifically for the health promotion program, they can enhance communications. Marquee announcements are best used in the same manner as public address announcements, for specific program reminders. They are often placed in cafeterias.

Electronic Mail

Electronic mail and electronic bulletin boards present a unique opportunity to promote health promotion activities in many settings. E-mail can be used for one-to-one communications, group distribution lists, and activity registrations. It can be used in lieu of memos especially in nationally dispersed organizations. General information can be shared on bulletin boards, but these require regular maintenance by health promotion staff.

PROMOTIONAL CAMPAIGNS

Promotional campaigns can be designed around a general health promotion theme or a variety of specific themes. A campaign can be defined as a series of promotional events that are planned around a single topic. Depending on the objective, the length of a campaign can be from a few weeks to several months.

In designing a promotional campaign, the first step is to establish a clear objective. This enables managers to plan activities that support the objective and to assess the campaign's effectiveness in achieving the objective. Promotional objectives vary:

- Increasing participation in major program activities
- Enhancing awareness of specific risk factors
- Disseminating information on a specific health topic
- Motivating specific changes in health behavior

A second consideration is the length of the campaign. To have a significant effect, the events of a health promotion campaign probably will need to be conducted over a period of several weeks. If there is strong interest in the theme of the campaign, or if each of the events is interesting and enjoyable in its own right, the campaign can be extended.

A third decision is the timing of the campaign events. There are many factors to consider:

- Proximity to a key event, such as a major health screening
- Proximity to related events taking place outside the organization, such as National Cancer Month
- Seasonal relevance, such as a "Spring Fitness Tune-up" campaign

In addition to the factors outlined above, managers also must consider the availability of financial and labor resources in planning a promotional campaign.

MANAGEMENT COMMUNICATION

Communication with management is an extremely important part of the promotion process. Health promotion communications must be developed specifically for managers. Timing, frequency, format, and level of detail are critical factors in an effective management communication program.

Timing

Management wants and needs to know what is going on at the workplace before employees know. In many organizations there are requirements for advance notice as well as an approval process. This enables management to determine how, if at all, the proposed activities will affect the organization's mission and objectives. This is another instance where solid relationships with senior managers will prove beneficial. It is also advisable to keep lower levels of management informed. Management support can make or break a health promotion program.

Frequency

The frequency with which health promotion managers communicate with site management will depend on the extent to which specific managers are involved with the program. Another factor is the importance of the messages. Although managers should be kept informed about program activities that affect them, care should be taken not to send unnecessary memos and announcements.

Format

Communications to management should be prepared in the format that is standard for the organization. This usually means a concise business writing style with little "hype." One-page memos on organization stationary are usually appropriate. For long documents, an "executive summary" format can be used. Managers also can make judicious use of promotional information as attachments or when the activity is specifically targeted to management. Managers are employees and therefore also should be program participants.

Level of Detail

The level of detail in management communications depends in part on the level of the manager being addressed. In addition, some managers want extremely detailed reports whereas others want to know only the key points. A way to meet the needs of the majority is to use an executive summary.

An executive summary is a one-page cover sheet that summarizes the essence of the communication. Supporting documents with more detail are attached. This allows individual managers to absorb as much detail as they want or need.

PUBLIC RELATIONS

Used appropriately, public relations efforts can be beneficial to the health promotion process. Public relations activities can be driven by forces that are either external or internal to the organization. Understanding these forces will enhance managers' effectiveness in promoting their programs.

An organization's public relations activities can be either proactive or reactive. Managers who maintain flexibility and some measure of spontaneity in programming and especially promotion can take advantage of reactive communication opportunities. For example, if a reporter and photographer arrive unexpectedly at the facility looking for photo opportunities in the human resources area, shots of happy employees participating in program activities can not only enhance the organization's image, but also cause management to view the program favorably.

In contrast, with advance notice from an organization's communications or public relations department, a health promotion manager can create a proactive communication opportunity. For example, an article on health care cost management can include remarks by the health promotion manager, the employee benefits manager, or the CEO that stress the organization's commitment to health promotion.

BRANDING

A major objective in promoting a product or service is to achieve some level of recognition among the consuming public. If the product or service is widely and well received, its brand name often becomes synonymous with all brands of the same product and becomes generic. This level of recognition often leads to increased market share because of the tendency for consumers who may know nothing about the product to buy the best-known brand.

The process of developing this identity is known as branding. In the field of health promotion, branding is another technique managers can use to promote their programs. Developing this brand identity involves identifying the program with the organization or its employees through the use of program names, logos, and a distinct "look."

Program Name

The name of a health promotion program may contain the sponsoring organization's name or some part of it. In other cases, names are more generic or identify the industry of which the organization is a part. The overall program name should be chosen carefully to convey a positive image and should be easy to remember without being faddish.

The names of specific activities can be health oriented or can reflect the organization's industry. For example, in a banking institution, "Well Bucks" could be the theme of an incentive program. An assembly operation could call an educational program "building a healthier you."

Program "Tag Line"

Along with the program name, a "tag line" is often used to further define the program. Examples are Johnson and Johnson Baby Shampoo, "The no more tears shampoo," and Gerber's Baby Food, "Babies are our business, our only business." A tag line can expand on a basic name and add another dimension to the product or service. Here are some examples of health promotion tag lines:

- "Our people count"
- "Good health is good business"
- "Where our employees are our greatest assets"
- "Helping achieve a healthier bottom line"

A tag line can communicate an organization's concern for employees, emphasize that the program supports business objectives, or convey any other desired image.

Consistent Program Themes

Themes can be developed for the entire program, the program for a year, or specific activities. Some program themes can be used in any situation:

- "Feel like a million"
- "Be the best you can be"
- "Go for the gold"
- "On top of the world"

Other themes that convey achievement, good health, or another positive image are equally effective. The theme should be used consistently in a variety of settings to unify the various aspects of an overall program. "Go for the gold" is obviously an Olympic theme. The program can feature Olympic-related activities, medals as incentives, or Olympic scoring. Participants who are pursuing fitness goals to win gold medals can relate to participants in other activities who also are working for medals.

Program Logo

The identity and image of a health promotion program can be promoted effectively with a distinctive logo that is used in conjunction with all program materials, communications, and incentives. The development of a logo requires care and thought because it will symbolize the program for a significant period of time.

A logo can be designed by a professional graphic artist who creates a concept and executes it. However, although the result may be excellent, the cost can be high. Also, the logo will not reflect participant input and the resulting identification with and commitment to the program. In contrast, using only participants' input can produce an ineffective and unappealing logo that will hurt the program's image.

A compromise approach is to begin the process by inviting participants to submit concepts and sketches. These can be screened by a committee and the best ones given to either an in-house or outside graphic artist. The artist can polish and enhance the submissions, and the committee can make its choice from among these. Once a logo has been selected, unveiling it publicly is a way to promote the program within the organization.

Program "Look"

The final step in developing a program identity is the creation of the overall program "look." Elements of the "look" are listed below:

- Colors of any printed materials
- Type style and size for official publications and communications
- A masthead for the newsletter
- Level of formality in printed communications: for example, use of a cartoon spokesperson rather than formal business-style announcements
- Type of paper used in printed materials: using recycled paper can enhance a program's healthy image

All the elements of program identity—name, tag line, theme, and logo—will be integrated into the overall look of the program. The result should portray the positive image with which both management and participants would like to identify. Images typically associated with health promotion programs are energy, vigor, quality, and stability.

Because health promotion programs are often targets of budget cuts in difficult times, it is wise not to use materials that convey a flashy or expensive image. However, neither should a program's look be allowed to become dated or stale. Often a simple change in a single component can freshen the program's image and attract new participants.

SUMMARY

The effective promotion of program activities first requires a thorough analysis of the communication systems and options in an organization. The next step is to select appropriate techniques to promote the program to different segments of the participant population. A well-coordinated promotional campaign is often one of the most successful of these techniques.

Another important component of promotion is branding. Consistency in the appearance of program materials and promotions can increase management commitment and employee participation and help differentiate the program from competitors.

This chapter concludes our discussion of marketing. Next we begin to consider the economic factors that affect health promotion program operations.

KEY TERMS

branding (p. 162)
communication meetings (p. 158)
program look (p. 163)
promotional campaign (p. 161)

public relations (p. 162)
tag line (p. 163)

SUGGESTED RESOURCES

Ailloni-Charas, D. (1984). *Promotion: A guide to effective promotional planning, strategies, and executions*. New York: Wiley.

Burnett, J. (1988). *Promotion management: A strategic approach*. St. Paul: West.

Hambleton, R. (1987). *The branding of America*. Dublin, NH: Yankee Books.

Lodish, L. (1986). *The advertising and promotion challenge: Vaguely right or precisely wrong*. New York: Oxford University Press.

Murphy, J. (1987). *Branding: A key marketing tool*. New York: McGraw-Hill.

Newsom, D., Scott, A., & Turk, J. (1989). *This is PR: The realities of public relations*. Belmont, CA: Wadsworth.

Shrimp, T., & DeLozier, M. (1986). *Promotion management and marketing communications*. Chicago: Dryden Press.

Part IV

Financial Operations

Our journey through the business disciplines has taken us from foundations and principles through management and marketing. We now address financial operations. Managers must know how to make the best use of limited financial resources to train and compensate employees and to develop, market, and sell products and services. In Part IV we seek to develop managers' understanding of the financial aspects of operating a health promotion department.

In chapter 18 we present a simplified picture of how the economy influences health promotion programs. We conclude with a discussion of how to prepare a program for hard economic times. In chapter 19 we present an overview of the budgeting process, including how to build and monitor a budget. The subject of chapter 20 is how to purchase products and services from outside vendors while following the organization's policies and procedures.

To this point we have covered three major areas of business: management, marketing, and budgeting. In order to make all of these functions work together toward a common goal a well-defined planning process must be used. In Part V our journey through the business disciplines ends with a discussion on business planning.

Chapter 18

Economic Influences

The financial condition of an organization and its economic environment determine how the organization functions. Conditions in the national economy and to an increasing extent the world economy influence the demand for an organization's goods and services, as well as the costs of labor, materials, and other aspects of doing business. The financial status of the organization also affects its policies and internal operations. To plan for the future and make appropriate decisions for the present, managers must be aware of economic trends and how they interact to affect the organization. In this chapter we discuss economic situations that affect health promotion and examine some economic indicators.

MACROECONOMICS

Macroeconomics is the study of the overall performance of the economy with respect to national production, consumption, average prices, and employment levels. Conditions in the overall economy significantly influence both for-profit and nonprofit organizations. In times of recession or depression, there is decreased demand for the products and services of for-profit businesses, and prices also decline. Unemployment rises and tax revenues fall, which decreases the funds available for government expenditures. Nonprofit organizations that rely on donations are likely to experience a decrease in contributions. Health promotion professionals should understand the economic conditions that contribute to these problems, as well as the indicators of economic trends.

Inflation and Recession

Two economic forces significantly influence the resources available to finance health promotion programs. These are inflation and recession. Inflation occurs when prices rise because the volume of money and credit is increasing faster than the amount of available goods and services.

A recession is a period of economic slowdown. Consumer confidence in the economy is low, and there is a decline in purchases of high-cost items such as houses, vehicles, and major appliances, even though their relative costs may be lower. Travel and other discretionary purchases also decline. As demand for goods and services falls, profits are reduced and businesses may lay off employees in an effort to reduce expenses.

Economic Indicators

Economies are dynamic. They have periods of growth and prosperity and periods of inflation or recession, and sometimes even depression. Because the health promotion field is affected by economic conditions, managers will find it worthwhile to monitor economic changes.

There are many indicators that provide insight into economic trends but none is completely predictive of the future. Some indicators for the U.S. economy are summarized in Figure 18.1.

MICROECONOMICS

Microeconomics is the study of the economic performance of the organization where decisions are a function of senior management's policies and actions. Health promotion managers can benefit from understanding the dynamics of these decisions and using the information to position themselves to avoid problems such as major budget cuts. Although some of these situations initially might appear threatening, they may actually benefit health promotion activities.

Mergers and Acquisitions

Mergers and acquisitions result in two organizations becoming one. In either case there are likely to be major changes in management policy and practice for at least one of the original organizations. Sometimes these changes result in an increase or a decrease in the number of employees who have access to health promotion services. The effect of a merger or acquisition on health promotion depends on the nature of the transaction and what kinds of health promotion services were offered previously.

A merger is usually an amicable transaction in which two organizations agree to combine operations. With regard to health promotion functions, a merger may be an opportunity or a potential problem because the new organization could end up with an expanded health promotion program or no program at all.

Health promotion managers in a potential merger situation should seek information about the transaction and make sure that senior managers are aware of the benefits of health promotion services and the advantages of continued program support.

In an acquisition, the acquiring organization has most or all of the power in deciding how the acquired organization will operate. In some cases the acquired organization continues to operate independently as a subsidiary of the new parent company; in other instances some or all of its operations are combined with those of the parent or with those of other subsidiaries.

The existence of a health promotion program in the acquiring organization may determine whether there will be a health

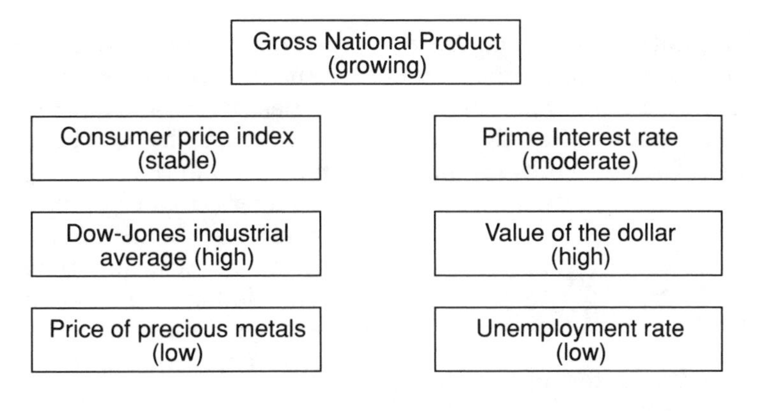

Figure 18.1 Health promotion programs generally fare best when the economic indicators reflect these trends.

promotion program in the acquired or combined organization. A health promotion manager in an acquiring organization should view the acquisition as an opportunity to expand the program to more employees. A manager in an acquired organization should promote the value of the existing program to the new management in any way feasible. An effective technique is to recruit program participants, especially those in powerful positions, to help lobby for continued and expanded health promotion programming.

Reorganization

To streamline its operations, increase profits, exploit new technologies, or enter new markets, an organization may restructure itself. This may result in a reassignment of duties, a decrease in the work force, or both.

A reorganization has both direct and indirect effects on health promotion. If there are personnel cuts, fewer employees will have access to the program. The program itself might be involved in the reorganization, changing the services offered or the number of health promotion professionals on staff. Whether the changes are positive or negative, the health promotion manager should play an active role in the reorganization to protect staff and services.

Some reorganizations result in a decentralization of operations from headquarters to various branch locations, each of which acquires more autonomy. Health promotion operations may be decentralized. Ideally, health promotion managers at branch locations would report to branch management, with major program development and purchasing continuing to be handled centrally. Local health promotion professionals can customize generic programs to meet their needs.

Sometimes an organization must reduce the number of its employees. This form of reorganization is called downsizing. Reasons an organization might need to downsize are

- loss of business to competitors,
- loss of business because of changes in consumer needs, and
- improved technology requiring fewer employees.

Whatever the reason, downsizing usually results in lower demand for health promotion services and could cause the loss of health promotion positions. When downsizing appears imminent, managers should market their programs more vigorously to avoid losing a large number of participants. This also will help forestall staff and budget cuts.

POSITIONING FOR HARD ECONOMIC TIMES

Understanding and monitoring economic trends enables managers to prepare effectively for internal and external changes that may affect the organization and the health promotion program.

As noted in chapter 4, it is important to integrate health promotion into other functions of the organization. If health promotion is an integral part of other departments, such as compensation and benefits, medical, employee assistance programs, occupational safety and health, training and development, risk management, sales, and public relations, it is more difficult for senior management to cut all or major parts of the program. This kind of integration takes a long time to develop and must be done on an ongoing basis.

Managers should maintain data on the program's utilization and success rates and have them organized and readily available when senior management requests the information.

Another way to avoid major program cuts is to charge at least a nominal fee for health promotion services. It is difficult for management to make major cuts when employees or others have paid for the services in advance. Also, it is advantageous to show management that the program is generating income.

SUMMARY

Health promotion professionals, like others in an organization, are subject to internal and external economic forces they cannot control. However, managers can be aware of these forces and plan for the changes they may bring. The first step is to monitor the industry and the economy. This will enable managers to develop positive responses to negative economic forces.

In mergers, acquisitions, and reorganizations managers also must monitor internal developments and plan to make changes positive for the health promotion program.

In addition to being aware of economic trends and forces, health promotion managers must develop, administer, and monitor the department's budget. The budgeting process is the topic of chapter 19.

KEY TERMS

acquisition (p. 170)
downsizing (p. 171)
inflation (p. 170)
merger (p. 170)
recession (p. 170)
reorganization (p. 171)

SUGGESTED RESOURCES

Feldstein, P. (1988). *Healthcare economics*. New York: Wiley.

Luke, R., & Bauer, J. (Eds.) (1982). *Issues in health economics*. Rockville, MD: Aspen.

Mair, D., & Miller, A. (Eds.) (1991). *A modern guide to economic thought*. Brookfield, VT: Elgar.

Sorkin, A. (1984). *Health economics, an introduction*. Lexington, MA: Lexington Books.

Chapter 19

Budgeting

Building, monitoring, and forecasting a budget is an important responsibility for health promotion managers. In this chapter we explain the need for budgets, describe some types of budgets, and outline the steps in developing a budget for a health promotion department.

THE NEED FOR BUDGETS

Budgeting gives senior managers the control they need to allocate resources and maintain a financially stable organization. At the level of the health promotion department, budgeting enables managers to ensure that the financial resources needed to achieve program goals are available.

TYPES OF BUDGETS

There are a number of ways to devise a budget. The format used depends on who develops the budget, how it is to be used, and whether the use of a particular format is mandated by an organization. A budget designed in one format can be translated into another format.

Two common budget formats are line item and functional. In this discussion we will emphasize the functional budget format. This format is easy for managers to use but may have to be translated into a line-item format. This results in two distinct formats for the same budget and will be referred to as a multiple-format budget.

Capital expenditures—the purchase of expensive durable goods—are not addressed in this text. These purchases are usually not made from a department's operating budget.

Line-Item Budget

A line-item budget shows specific items and the costs associated with them. Typical line items found in a health promotion environment are

- labor;
- materials;
- supplies;
- travel;
- purchased services;
- shipping and freight;
- telephones;
- occupancy, rent, or lease;
- overhead; and
- allocations for centrally administered programs, such as health promotions.

Some of these categories can be broken down into more specific classifications. For example, labor can be divided into full time, part time, exempt, or nonexempt. Telephone expense can be divided into fixed costs, local charges, and long-distance charges. Most accounting departments use a number as well as a descriptor for each category. These are usually referred to as account numbers or accounts. A basic line-item budget organized as described above is shown in Figure 19.1.

Functional Budget

A functional budget is based on the functional areas of a typical program. Specific line items are included in each area. Some functional areas in health promotion programs are

- orientation,
- health screening,
- education,
- group activities,
- incentive programs,
- program evaluation,
- promotion, and
- fitness center.

The advantage of a functional budget in a health promotion environment is that it can be directly related to an annual program plan. This allows managers to monitor the costs for any specific activity area that is included as a functional category in the budget. Any of the line items described above can be included in the functional areas.

Multiple-Format Budget

Although financial managers may not specify the method used to prepare department budgets, they may want the finished product in a line-item format. For this reason, managers who use a functional budget may have to maintain a multiple-format budget. This is fairly easy when a computer spreadsheet is used. An account summary showing line-item expenses was extracted from a functional budget and is shown in Figure 19.1.

BUILDING A BUDGET

The budget process begins with department managers submitting their requests for funding. The best way to determine what those requests should be is to build a budget. In health promotion settings, managers can begin this process by gathering data about all costs and sources of revenue.

Data Gathering

Before attempting to develop a budget, managers should gather as much relevant data as possible. Like preparing an income tax return, the process goes more smoothly if all the information is at hand.

ACCOUNT SUMMARY

Site designation:	NYC
For the budget period:	199X
Prepared by:	M. E. Manager
Date of preparation:	199X

Account Number	Total
2005 Clerical Labor	$ 8,840
2010 Administrative Labor	$33,252
2020 Management/Consultant Labor	$ 2,075
3050 Program Materials	$ 1,856
3110 Printing—Outside	$0
3112 Electronic Mail	$ 180
3120 Printing—Internal	$ 1,360
3124 Xerox—Internal	$ 1,400
3130 Office Supplies	$ 300
4550 Telephone—Fixed	$ 1,200
4720 Maintenance	$ 100
5010 Professional Services (Purchased Services)	$876
5030 Consultants (Purchased Services)	$0
6005 Fringe Benefits	$ 8,833
6270 Seminars & Conferences	$0
6510 Travel	$1,384
7250 Freight—Surface	$ 100
7910 Sundry	$ 1,660
8010 Occupancy	$15,000
8024 Word Processing	$0
8260 Long Distance	$ 1,200
8261 Telephone Fixed—In	$ 2,400
8272 Postage	$ 240
Total Expense	$82,256

Figure 19.1 A line-item budget is used to organize revenue and expenses.

The key areas in which data are needed are the eligible population, the quantified program goals, and the costs of materials and labor. These three factors account for over 50% of the line items in a health promotion budget. It is possible to construct a spreadsheet with formulas that will automatically calculate these line items.

Eligible Population

A basic piece of information needed for budgeting is an accurate estimate of the program's eligible population. This may be the expected high figure for the budget year or an average figure. This information is usually available from the human resources department.

Eligibility is determined by the department that is responsible for funding the program. In corporations this is often the employee benefits department. In any setting, managers should be sure to include all eligible participants. These may include

- full-time employees,
- part-time employees,
- retirees,
- dependents,
- all or some union members depending on whether health promotion is a negotiated benefit,
- certain divisions or departments,
- participants in certain geographic areas, and
- residents of certain areas for programs delivered by public health agencies.

Program Goals

The second important piece of information is the specific program goals the budget is expected to fund. Obviously, when program goals are multiplied by the eligible population, the actual number of participants that will be consuming materials or requiring labor will be determined. These are examples of program goals:

- 20 percent of eligible population participating in stress management classes
- 45 percent of eligible population completing fitness activity with incentives
- 80 percent of eligible dependents participating in health screening

Costs

The third area where preliminary data should be gathered is predictable costs. These may be fixed or variable (related to participation). The largest variable-cost items are labor, materials, and services.

Labor. Labor costs can be fixed or variable. Fixed costs include

- salaries for permanent full-time and part-time employees, and
- wages for part-time employees who are used for activities such as health screening where the program is committed to a certain minimum number of hours.

Variable costs include wages for part-time employees whose services are used in proportion to participation in activities such as

- education classes,
- fitness assessments by appointment, and
- part-time staffing for health screening where the number of hours worked depends on the level of participation.

Average hourly wage figures will be sufficient for building the budget. Payments for contracted services are not considered direct labor costs because these services are provided by vendors and as such will probably be placed in a line item such as purchased services, professional services, or contracted services.

Materials. In some cases costs for materials can be stated accurately because a purchasing decision has been made. In other instances a fairly accurate estimate can be made based on previous experience or a range of bids. Examples of materials used in health promotion programs are

- textbooks and handouts,
- incentive items such as T-shirts,
- health risk assessment questionnaires, answer sheets, and output documents,

- refreshments served at health screening, and
- newsletters.

Services. As with materials, the costs of services can be stated accurately or projected. Typical health promotion services may include

- contracted services such as body composition testing,
- cholesterol testing,
- contracted instructors for classes, and
- health risk assessment processing costs.

Other cost items are mileage expenses, benefits percentages, burden rates, and costs that may be unique to the organization.

Calculating Costs

The next step in the budget building process is to calculate all variable and fixed costs. Most of the costs associated with health promotion are variable. These can be calculated by a computer spreadsheet program. Fixed costs must be calculated or recorded manually.

Variable Costs

In a health promotion program, variable costs are costs that vary with the level of participation. For each person who participates in an activity, the cost for the activity increases by one unit. For example, if the cost of a health risk assessment is $9.45 per individual, then for each individual who participates, the cost will increase by $9.45. This holds true for all totally variable costs. When costs are mixed or semivariable, an economy of scale may enter the equation as participation increases.

The calculation of variable costs based on the information gathered about population, goals, and costs will be illustrated by using portions of a computer spreadsheet that was designed by the author in which 51% of the line items were calculated through either simple formulas or ratios. Many of these items were covered in the preceding discussion of costs. A list of program goals that will be used in the calculation of some variable costs follows.

Figure 19.2 shows a list of costs that are common to many health promotion programs.

The result of the relationship between population size and goals is built into the

PROGRAM GOALS

Enter goals or related data	
Employee Population	0
Expected New Employees	0
Eligible Spouses	0
Enter goals as a percentage of eligible populations	
% of Employee Enrollment	0%
% of Spouse Enrollment	0%
Health Risk Appraisal Screening	
% of Employees Screened	0%
% of Spouses Screened	0%
Education	
% in Education (total)	0%
% in Instructor-Led	0%
% in Weight Control	0%
% in Nutrition	0%
% in Stress Management	0%
% in Fitness	0%
% in Smoking Cessation	0%
% in Back Care	0%
% in Other (Specify Topic Below)	0%
% in Self-Study	0%
% in Weight Control	0%
% in Nutrition	0%
% in Stress Management	0%
% in Fitness	0%
% in Smoking Cessation	0%

(continued)

(continued)

Course Completion Rates	
Instructor-Led (%)	0%
Self-Study (%)	0%
Action Teams	
% in Action Teams	0%
Special Events/Seminars	
% in Special Events or Seminars	0%
% of eligible population at serious risk in relevant activities	
Exercise	0%
Weight	0%
Smoking	0%
Hypertension	0%
Cholesterol	0%
Seat Belts	0%
Back Pain	0%

section of the spreadsheet shown in Figure 19.3. An illustration of a typical formula for total health assessment follows.

Assuming:

Total employee population is 1,000.

50% of the population has a spouse.

Program enrollment is 80% for employee and 20% for spouses.

The health assessment goal is 50% of the enrolled population.

This would be the calculation for total health assessment:

Total population of 1,000 times an enrollment goal of 80% yields 800 enrolled employees.

Total enrolled spouse population is 50% times 1,000 employees times an enrollment goal of 20% yields 100 enrolled spouses.

800 enrolled employees plus 100 enrolled spouses times a health assessment goal of 50% yields 450 enrolled individuals participating in health assessment screening.

All of the calculations that comprise the formulas in Figure 19.3 are derived with similar logic.

These participation figures are then multiplied by per-unit costs to determine variable costs. All of the results used in the materials section of Figure 19.4 relate the participant numbers from Figure 19.3 and the per-unit costs from Figure 19.2. Organizing variable costs in this manner on a computer spreadsheet allows managers to make changes in population, goals, or costs and see the resulting budget impact in seconds.

Fixed Costs

Fixed costs are costs that do not vary with participation. They are incurred whether one person or 1,000 participate. Labor is an excellent illustration of the difference between fixed and variable costs. The labor section of Figure 19.4 has both fixed and variable line items. The salaries of permanent employees such as the manager and support staffers are fixed costs. However, labor that relates to education or other program activities if part-time employees are used tends to be variable.

In the Supplies section of Figure 19.4, the costs of many supply items can be determined by formulas as illustrated above. However, once again there are exceptions, such as the major startup supplies category. For new programs this could include an initial basic stock of stationery, desk accessories, or other one-time major purchases. Later the cost to replenish these supplies can be approximated with ratios based on the first year's usage. The line items in Figure 19.4 serve as further illustrations of fixed costs.

COST DATA ENTRY LINE ITEMS

Enter average fixed costs per unit, etc.		
Program Introduction Mailings	$0.00	
Orientation Packets	$0.00	
HRA Questionnaire	$0.00	
HRA Printout Materials	$0.00	
HRA Processing	$0.00	
Blood Drawing Labor (per sample)	$0.00	
Blood Processing (per sample)	$0.00	
Temporary Clerical Hourly Labor Rate	$0.00	
Screening Rate (Participants/Hour)	0	Average
Number of Clerical Temps at Screening	0	
Screening Refreshments (per person)	$0.00	
Instructor-Led Participant Kits	$0.00	Average
Weight Control	$0.00	
Nutrition	$0.00	
Stress Management	$0.00	
Fitness	$0.00	
Smoking Cessation	$0.00	
Back Care	$0.00	
Other (enter average for all topics)	$0.00	
Self-Study Participant Kits	$0.00	Average
Weight Control	$0.00	
Nutrition	$0.00	
Stress Management	$0.00	
Fitness	$0.00	
Smoking Cessation	$0.00	
Campaign Booklets	$0.00	Average
Best Buys	$0.00	
Defensive Driving	$0.00	
Drinking and Driving	$0.00	
Seat Belts	$0.00	
Average Instructor-Led Class Size	0	
Average Instructor Wage (per hour)	$0.00	
Do you contract for course instructors?		
Enter 1 for yes or 2 for no	0	
Average Contract Amount per Person	$0.00	
One-Way Instructor Commute (miles)	0	
Mileage Reimbursement Rate	$0.00	
Action Team Incentive	$0.00	
Course Completion Incentive	$0.00	
Newsletter Subscription Rate	$0.00	
Average Cost per Page for Copying	$0.00	
Newsletter Cost per Individual Issue	$0.00	
Number of Issues per Year	0	
Current Labor Burden Rate (overhead %)	0%	

Figure 19.2 Costs associated with health promotion programs.

GOAL AND POPULATION FORMULA RESULTS

Employee Enrollment	0
Spouse Enrollment	0
Total Enrollment	0
Employees in Health Screening	0
Spouses in Health Screening	0
Total Health Screening	0
Education Total	0
Instructor-Led Total	0
Instructor-Led Completions	0
Total Number of Instructor-Led Classes	0
Self-Study Total	0
Self-Study Completions	0
Action Team Participants	0
Special Event/Seminar Participants	0

Figure 19.3 A population and goals interaction spreadsheet helps managers determine variable costs.

Mixed Costs

Mixed costs have variable and fixed components. During the budget building process, it is generally easier to simply leave a mixed-cost item in the category in which it is probably going to remain most of the time. However, some items are truly mixed, such as the electronic mail item in the phone section of Figure 19.4. In this case there is a fixed monthly charge for the basic service as well as a variable on-line time charge. This also may be true of long-distance phone rates or other expenses. In a functional budget as illustrated by Figures 19.1 through 19.4, it is easy to cover all cases.

The example used to illustrate the process of identifying fixed and variable costs was taken from a real-world health promotion program. However, it does not address all possible situations, such as the costs of operating a fitness center. Also, as noted earlier, this text does not address capital expenditures or depreciation. The key point is that during the budget-building process, all identifiable costs should be included.

Sources of Revenue

Many health promotion programs are funded entirely by the sponsoring organization and have no outside sources of revenue. In other settings the program is partially funded by outside sources, such as

- participant copayments,
- user fees for specific activities,
- proceeds from the sale of program materials, and
- proceeds (or a percentage) from recreation club dues or company store sales.

In these partially funded programs, revenue figures must be accurately estimated because the organization must provide the remaining funding. If outside sources do not produce the expected revenues, it may be necessary to implement contingency plans to make up the shortfall.

Leveraging Resources

Leveraging resources is the process of accessing other sources of program funding from both within and outside the organization. Another department may provide a cash subsidy or be willing to perform a service such as producing a newsletter in exchange for special health promotion services. Leveraging resources is one of the most challenging tasks managers can accomplish. The opportunities and techniques are limited only by the imagination.

Some leveraging methods are

- obtaining clerical support from other departments without being charged for it,
- using participant volunteers,
- obtaining incentive items from the marketing department,

VARIABLE COST SPREADSHEET

Labor and Material Line Items

Labor	***Fixed***	***Variable***	
Enter annual salaries			
Project Manager/Consultant(s)	$0		$0
Project Administrator/Intern(s)	$0		$0
Clerical(s)	$0		$0
Education (Instructor-Led)		$0	$0
Instructor Usage (other than courses)		$0	$0
Labor Subtotal	$0	$0	$0
Total Burden/Overhead	$0	$0	$0
Total Burdened Labor	$0	$0	$0

Materials	***Fixed***	***Variable***	***Total***
Program Introduction Mailings		$0	$0
Orientation		$0	$0
HRA Questionnaires		$0	$0
HRA Printouts		$0	$0
Instructor-Led Participant Kits		$0	$0
Self-Study Participant Kits		$0	$0
Newsletter		$0	$0
Brochures (other than orientation)		$0	$0
Large Posters		$0	$0
Small Posters		$0	$0
Table Tents		$0	$0
Health Highlights		$0	$0
Campaign Guides		$0	$0
Campaign Booklets		$0	$0
Miscellaneous Promotion		$0	$0
Other: specify below and enter			
		$0	$0
		$0	$0
		$0	$0
Total Materials	**$0**	**$0**	**$0**

(continued)

Figure 19.4 The variable cost spreadsheet helps managers make necessary budget changes needed as a result of changes in the populations, goals, or costs.

Supplies, Travel, and Purchased Services Line Items

Supplies	***Fixed***	***Variable***	***Total***
Major Startup Supplies	$0		$0
Office		$0	$0
Postage		$0	$0
HRA Screening Refreshments		$0	$0
Course Completion Incentives		$0	$0
Action Team Incentives		$0	$0
Promotional Supplies		$0	$0
Special Events Supplies		$0	$0
Other: specify below and enter			
		$0	$0
		$0	$0
Total Supplies	$0	$0	$0

Travel	***Fixed***	***Variable***	***Total***
Staff (routine)		$0	$0
Staff (project management)		$0	$0
Education		$0	$0
Area or National Meeting(s)	$0		$0
Other: specify below and enter			
		$0	$0
		$0	$0
		$0	$0
Total Travel	**$0**	**$0**	**$0**

Purchased Services	***Fixed***	***Variable***	***Total***
Blood and/or HRA Processing		$0	$0
HRA Screening Medical Staff		$0	$0
HRA Screening Clerical Labor		$0	$0
Service/Repair		$0	$0
External Printing/Pubs and Graphics		$0	$0
Other Temporary Clerical Labor		$0	$0
Equipment Leasing	$0		$0
Instructor-Led Education Contracts		$0	$0
Other Subcontracted Services	$0		$0
Other: specify below and enter			
		$0	$0
		$0	$0
		$0	$0
Total Purchased Services	**$0**	**$0**	**$0**

Figure 19.4 *(continued)*

Shipping, Freight, Phones, Occupancy, and Miscellaneous Line Items

Shipping/Freight	***Fixed***	***Variable***	***Total***
Other: specify below and enter			
		$0	$0
		$0	$0
		$0	$0
Total Shipping/Freight	**$0**	**$0**	**$0**

Phones	***Fixed***	***Variable***	***Total***
Equipment Charges	$0		$0
Local Service Charges		$0	$0
Long Distance Service		$0	$0
Electronic Mail	$0	$0	$0
Other: specify below and enter			
		$0	$0
		$0	$0
Total Phones	**$0**	**$0**	**$0**

Occupancy/Rent/Lease	***Fixed***	***Variable***	***Total***
Annual Charges	$0		$0
Other: specify below and enter			
		$0	$0
		$0	$0
Total Occupancy/Rent/Lease	**$0**	**$0**	**$0**

Other	***Fixed***	***Variable***	***Total***
Word Processing		$0	$0
Internal Printing/Copying		$0	$0
Internal Pubs and Graphics		$0	$0
Newsletter Publication		$0	$0
Staff Development	$0		$0
Other: specify below and enter			
		$0	$0
		$0	$0
		$0	$0
Total Other	**$0**	**$0**	**$0**

	Fixed	Variable	Total
Project Totals	$0	$0	$0
Cost per Eligible Employee	$0	$0	$0

Figure 19.4 *(continued)*

- using local vendors as speakers in exchange for the right to distribute sales materials,
- accepting donations of prizes for health fair drawings from local merchants,
- using local facilities such as a school gymnasium for sports or fitness activities, and
- using public agency materials or volunteers.

Putting the Budget Together

After all costs and all sources of revenue have been considered, the budget can be assembled in a format that meets the needs of both health promotion managers and financial managers. A key indicator that can be used to compare the efficiency of a program from year to year is a cost per eligible figure. This figure is the total budget amount divided by the total eligible population. Other figures that can be useful are

- cost per enrolled participant,
- fixed cost per participant, and
- variable cost per participant.

Managers should strive to drive down the cost per participant without affecting program quality. One way to do this is to increase participation in all aspects of the program, particularly in low-variable cost activities such as walking or running clubs that may require only occasional incentives. This has the effect of spreading the fixed costs over a larger population and reducing the cost per participant.

Another approach is actual cost cutting. This may involve reducing staff, eliminating low-participation/special-interest program offerings, or reducing hours for fitness centers. Some of these measures should be seriously discussed before implementation because they may reduce program quality.

Table 19.1
Top-Down and Bottom-Up Budgeting: Advantages and Disadvantges

Method	Advantages	Disadvantages
Top down	Expedient Not labor intensive	Less commitment by line managers Less accuracy
Bottom up	High accuracy Strong commitment by line managers	Time consuming Labor intensive

The final stage of developing a budget involves spreading the budget figures over a 12-month period. This is done by referring to the program plan, which should specify the months in which various programs will take place. Some fixed items such as salaries will be spread out evenly over the year. Other allocations will be for a specific period, such as a Cancer Month activity. Data from past years may show that certain expenses tend to accumulate in the fourth quarter. This trend should be considered in allocating expenditures over the budget period.

BUDGET DEVELOPMENT PROCESSES

The budget development process varies from one organization to another. In most cases it involves a number of iterations. Two approaches are a top-down process and a bottom-up process. The advantages and disadvantages of each are listed in Table 19.1.

Top-Down Budgeting

In top-down budgeting, senior managers determine a suggested budget amount for each department or other operating unit. In most cases this initial budget figure is flexible, and lower levels of management can negotiate for additional funds by providing new information to the decision makers.

Bottom-Up Budgeting

Although it involves more work, most managers prefer a bottom-up budgeting procedure. Each manager must submit a preliminary budget well before the end of the fiscal year. Senior management provides guidelines and allocations to overhead along with a deadline for completion. Managers then follow the procedures described in this chapter and develop a budget request that supports the department's long-range plan.

Decision makers review each budget request in the context of the overall budget and the submissions of other departments. Senior managers suggest changes, challenge assumptions, and in some cases order cuts in each budget. The manager then reworks the budget and resubmits it, at which point it is either approved or sent back for more revisions. The process continues until final budget figures are approved.

TRACKING AND MONITORING BUDGETS

After budgets have been approved and the new fiscal year has begun, managers will be asked to track their budgets and monitor expenses and revenues. This process is the foundation for forecasting, which is critical to the success of the organization's plans and objectives.

The Relationship of Forecasting

Accurate tracking of a budget gives managers the information they need to make adjustments in spending and program planning and to accurately forecast results for senior management. This information, which begins at the micro level within an individual department, is rolled up into the organization's overall budget forecast. Financial managers use the forecast as a basis for decisions about cash flow, debt financing, and labor reductions. Because many of these decisions have far-reaching effects, it is important that department forecasts be as accurate as possible.

Forecasting Budgets

In many organizations, managers must review each month's budget printout for accuracy and make changes in the budget forecast based on year-to-date experience, program plan, changes in assumptions, and any other new information. There are two basic methods of forecasting budgets: straight-line forecasting (also known as "run-rate" forecasting) and forecasting by plan. At the end of the fiscal year, forecasting accruals becomes especially important. Accruals are expenses that are identified as having been incurred. Funds then are set aside for future payment.

Straight-line forecasting is based on average expenditures for a previous period of time. This process works well when there is little variation from one budget period to another. The accuracy of a straight-line forecast increases as the year progresses. It would be inappropriate to use straight-line forecasting until at least three or four months of experience have been monitored because a single month's variation can skew the average. Even then, the decision to use this method should be pondered if there are significant fluctuations from month to month. After six months, it becomes more reliable. Straight-line forecasts are also used when there is little information about the future or an extremely short period of time to calculate the forecast.

A preferred method of forecasting is one that is done in direct relation to the budget and program plan. In this method, the year-to-date history is seldom averaged. It is reviewed in relation to goals.

Forecasting by plan requires a detailed analysis of the results of the monitoring process and thorough familiarity with the budget itself. It compares actual line-item expenses with forecasted expenses and actual participation with goals. Adjustments are made in the budget forecast based on adjustments to specific future activity goals. It also requires highly developed planning skills, including contingency planning. This method works well when the bottom-up process was used to develop the budget and the manager has authority to make both program and budget decisions. It results in a more accurate forecast than the straight-line method.

The Monitoring Process

To make accurate forecasts, it is necessary to monitor experience carefully. This requires

- accurate records, such as copies of invoices and purchase orders, or detailed logs;
- easily retrievable information; and
- well-organized information, usually by date and either functional area, accounting line item, or both.

Here again, computer spreadsheets are an invaluable tool in all aspects of budgeting, including monitoring.

SUMMARY

Through the budgeting process, managers can significantly improve program quality and help the organization achieve its goals. The ability to develop a detailed functional budget from the bottom up will enable managers to monitor experience and prepare accurate budget forecasts throughout the year. This will provide the best opportunity to achieve program goals.

In this chapter we considered budgeting as a means of allocating funds for the purchase of goods and services. In chapter 20 we examine the actual process of purchasing.

KEY TERMS

accrual (p. 186)
bottom-up budgeting (p. 185)
budget (p. 173)
forecasting (p. 186)
fixed costs (p. 180)
leveraging (p. 184)
mixed costs (p. 180)
top-down budgeting (p. 185)
variable costs (p. 177)

SUGGESTED RESOURCES

Dawson, J. (1985). *A model for systematic budgeting*. Santa Monica, CA: Rand.

Donnelly, R. (1984). *Guidebook to planning: Strategic planning and budgeting basics for the growing firm*. New York: Van Nostrand Reinhold.

Herkimer, A. (1988). *Understanding health care budgeting*. Rockville, MD: Aspen.

Koven, S. (1988). *Ideological budgeting: The influence of political philosophy on public policy*. New York: Praeger.

Sweeney, H., & Rachlin, R. (1987). *Handbook of budgeting*. New York: Wiley.

Wildavsky, A. (1986). *A comparative theory of budgetary processes*. New Brunswick, NJ: Transaction Books.

Chapter 20

Purchasing

No health promotion program is entirely self-produced. Managers must rely on external sources for a wide array of raw materials or finished products. These resources are obtained through the purchasing process. This may be a highly structured process administered through a purchasing department, or it may be a more informal process handled entirely by the health promotion manager.

A TYPICAL PURCHASING PROCESS

In most large organizations, there is a standard purchasing process. In smaller organizations, managers may follow the same procedures in a less structured and rigid manner. The typical purchasing process involves the use of purchase requisitions and purchase orders. There is a defined vendor selection process and a procedure for handling vendor relations. These are discussed below.

Purchase Requisitions

A purchase requisition is a document that is used to begin the purchasing process. It is completed by the person who wishes to make the purchase and is designed to describe accurately the requirements for the materials or services to be purchased. It also includes information that allows the accounting department to charge the purchase to the correct division, department, and account. The completed requisition gives the purchasing department the information it needs to obtain competitive bids and make a purchasing decision. A sample purchase requisition is shown in Figure 20.1.

This is the information that is typically needed for a purchase requisition:

- Requisitioner identification, including division, department, and account numbers
- The requisitioner's address
- The address to which the materials or services should be delivered
- The address to which the vendor should send the invoice for payment
- A detailed description of the product or service
- The quantity of the product to be purchased
- A suggested vendor if one is known
- The date by which the purchase must be delivered to the requisitioner

The requisition also will require a series of signoffs to authorize the purchase. These usually include the requisitioner and one or more higher levels of management. Approval at still higher levels may be required when the cost of an item exceeds a certain amount. The completed purchase requisition is sent to the purchasing department, manager, or agent.

Purchase Orders

On receiving a completed purchase requisition, the purchasing department proceeds to make a purchase decision. This may be the result of competitive bidding, an organization-wide volume purchasing agreement, or some other method. The purchase decision is communicated to the vendor through the use of a purchase order. A sample purchase order is shown in Figure 20.2.

A purchase order contains most of the information provided on the requisition and also specifies the exact product to be purchased, the manufacturer, and the vendor of choice. It is signed by an employee who can authorize purchases.

Most vendors will not deliver products without an authorized purchase order. They will not accept a purchase requisition. The reason for this is that many organizations will not pay invoices unless they contain an authorized purchase order number.

Selection of Vendors

In selecting vendors, there are specific steps a requisitioner can take to make the process go more smoothly. Health promotion managers as requisitioners can assist the purchasing department in three ways: doing some nonthreatening preliminary research, making specific product or vendor recommendations, and educating the purchasing department about the department's needs and how it operates.

To aid in conducting preliminary research, it is useful to maintain current files of vendor literature and catalogues for materials, supplies, and services that may be needed in the course of program planning. Another way to research vendors is by word-of-mouth recommendations from an active network of professional colleagues. This is a valuable source of information because the network members usually have no vested interest in vendors and can make unbiased statements about the practices of vendors and the quality of their products and services.

Although it is appropriate to question prospective vendors, managers should be careful not to undermine the functions of the purchasing department. The information a manager gathers can be passed to the purchasing department in a nonthreatening manner that will help build a productive working relationship with this important department.

PURCHASE REQUISITION

Requisition no. 535987

Page of	Requisitioner	**Preparer**: Complete the cost distribution information according to the accounting system used in your organization.
Telephone	Facility code	

Div.	Dept.	Account	GLA	Project/CEA

Suggested vendor	Date of request
	Phone

Street address

City/State/ZIP code

Attention: ❑ Confirming order placed with: ____ Date ____ ❑ Non confirming

Deliver to address

Company name

Street address

City/State/ZIP code

Attention: (Name, facility code)

Invoice to address

Commodity	Terms	F.O.B.
	_._%___ N ___	

√	Change	Type	Savings	Ship via: 1=Surf 2=Air 3=Vendor

TAX CODE: State | County | City

❑ Taxable ❑ Non taxable ❑ Partial

Item no.	SUF	Quantity	U/M	Drawing/part no.	Rev	Insp	Complete description of goods or services	Required delivery	Unit price	D	Extended price

Requisition approval*	Must be within currently approved delegation of authority for the total estimated cost shown on this requisition				*Total estimated cost	Total order	
Name & title							
Signature & date							

Figure 20.1 A purchase requisition begins the purchasing process.

PURCHASE ORDER

Page of	Requisitioner
Telephone	Facility code

PO #
535987

Div.	Dept.	Account	GLA	Project/CEA

	Date of request
	Phone
Vendor number	
Selected vendor	
Street address	
City/State/ZIP code	
Attention:	

Deliver to address
Company name
Street address
City/State/ZIP code
Attention: (Name, facility code)

Invoice to address

Purchase order no.			Commodity	Terms		F.O.B.
Div.	Buyer	Order no.		. % N		
			√ Change	Type	Savings	Ship via: 1=Surf 2=Air 3=Vendor
TAX CODE State	County	City	❑ Taxable ❑ Non taxable ❑ Partial			

Item no.	SUF	Quantity	U/M	Drawing/part no.	Rev	Insp	Complete description of goods or services	Required delivery	Unit price	D	Extended price

Total estimated cost	Total order		
	Purchase order approval		
	Buyer	Manager	

Figure 20.2 The purchase order represents the agreement between the organization and the vendor.

A requisitioner also can assist in the purchasing process by making specific recommendations to the purchasing department. These recommendations should be based on research findings or firsthand knowledge of the product and vendor.

Particularly when a health promotion program is new in an organization, the manager can help the purchasing department by educating its employees about the program's specific needs. This will enable the purchasing employees to do their jobs more efficiently and save the health promotion manager time and effort over the long term.

This effort can begin with a meeting where the manager describes the general needs of the department and gives catalogues and business cards of vendors to purchasing employees. The education process can continue on a purchase-by-purchase basis as the program expands.

Vendor Relations

Many organizations have established procedures with respect to vendor relations. These procedures are used to resolve conflicts caused by miscommunications, late payment of invoices, product defects, or other sources of dissatisfaction for either the requisitioner or the vendor.

Vendor relations are important for several reasons. First, if a specific vendor is the only one that can provide a product or service, that vendor has the power to disrupt program activities by withholding shipments of needed materials because of an unresolved issue. Second, problems with vendors can lead to litigation; this can be avoided by establishing a productive vendor relations procedure. Finally, managers must be familiar with any organizational policies that deal with vendor relations and conflicts of interest.

PURCHASING SERVICES

In some organizations, the procedure for purchasing services may be identical to that for purchasing products. Other organizations may require additional paperwork. Also, the vendor selection process may be different because it is difficult to quantify or qualify services.

Consulting Agreements

Many services fall into the category of consulting. Consulting is a broadly defined term that can include any of the following:

- Individuals who actually consult, that is, give advice and recommendations
- Individuals who offer specific services under the auspices of their own business, such as an independent aerobics instructor
- Small businesses that provide services such as maintenance, temporary labor, or graphics work

In many organizations, a consulting agreement must be executed to obtain these services. The consulting agreement is a document that contains these elements:

- Specific information about the proposed vendor, such as name, address, and telephone number
- Disclaimers, waivers, and other clauses that limit the organization's legal liability
- A specific description of the services to be delivered
- A statement of the method, amount, and timing of compensation
- Specifics regarding deadlines and the length of the agreement

The agreement may contain other elements that are specific to the organization.

Selecting Service Providers

The process of selecting service providers is less exact than that of selecting product vendors because of the intangible nature of services. For this reason it is important to talk directly with other health promotion managers who have used a provider for similar services. Ask for specifics about the quality of the services as well as the people who provide them.

When evaluating these recommendations, it is important to identify the nature and patterns of mistakes. However, it is unrealistic to expect a vendor to perform flawlessly all the time. When vendors make mistakes, the customer should consider how they responded to the situation. Did they assume responsibility? Did they make extraordinary efforts to correct the mistake? Did they offer compensation such as a price reduction? Did they respond promptly and treat the customer with respect? If vendors handle errors appropriately, occasional minor mistakes may be overlooked. Of course, it is necessary to reevaluate a relationship with a vendor if there appears to be a pattern of frequent and/or serious errors.

It also may be useful to have several staff members interview prospective vendors in a process similar to group interviewing of prospective employees. The group can then rank the vendors and make a recommendation to the purchasing department. Before undertaking this effort, managers should be sure they are not usurping the authority of the purchasing department.

Requests for Proposal

Another method of selecting vendors is the request for proposal (RFP) process. This is usually used for large projects that involve complex services. An RFP may be appropriate in selecting vendors to

- manage a fitness center,
- manage an employee assistance program,
- conduct the entire health screening operation, or
- manage the entire health promotion program.

After writing an RFP, the manager sends it to the vendors on the list of bidders, evaluates the proposals, decides on a "short list," interviews the finalists, and lets the bid. Most organizations have an established procedure for RFPs, and managers should be familiar with it.

OTHER PURCHASING CONSIDERATIONS

There are several additional aspects of purchasing managers must consider. They are statements of work, negotiating, sole-source purchasing, and the role of the legal department.

Writing Statements of Work

The ability to write clear, concise, and detailed descriptions of your expectations of a vendor is a valuable skill. Such descriptions are called statements of work and are needed for purchase requisitions, consulting agreements, requests for proposals, and contracts for services. If correctly written they will allow vendors to accurately bid on projects and deliver services efficiently. They also will enable managers to identify improperly or inadequately delivered services and either seek a remedy from the vendor or dismiss the vendor.

Statements of work should contain the following elements:

- A description of the nature of the work, for example, instruction, administration of tests, evaluation of test results

- A description of the time frame within which the work is to be done
- Clear descriptions of the responsibilities of the vendor as well as the organization, if appropriate
- Reporting requirements if needed

An example of a statement of work appears on this page. Although a statement need not be this long and detailed, there should be no doubt as to who does what and when.

Negotiating Price, Terms, and Conditions

Negotiating the various aspects of an agreement for services is worthy of an entire course in itself. Indeed, full treatment of this subject is beyond the scope of this text. Students should consult the list of suggested resources at the end of the chapter for additional information. A brief overview of the role and importance of negotiating follows.

The purpose of negotiating is to amicably reach an agreement that meets the needs of both parties. This is often referred to as win-win negotiations. A clear sign that negotiations were successful is that both parties not only believe that their needs were met, but also would willingly sit down with the other party in the future to negotiate again.

The negotiating process may involve several long discussions that call for creativity and flexibility on the part of both parties to reach a win-win situation.

Here are some suggestions for negotiating effectively:

- Negotiators should enter the negotiation with a clear picture of their desired outcome.

- Negotiators should know what will be their final offer before the negotiations begin.

- Negotiators should have an idea of the needs of the other party.

- Negotiators should have a sincere desire to reach an agreement.

- Negotiators should know that they can always choose to walk away from the negotiations. If they do not have that ability,

Statement of Work

The vendor will offer stress management classes at the headquarters building twice weekly during the current calendar year. The classes will consist of lecture and discussion and be 1 hour in length. Each offering will consist of a series of 6 classes over a 3-week period. The instructor will hand out evaluation forms at the end of each 3-week session and return them to the program administrator within 1 week of the end of the session. The course outline will be mutually agreed on before the commencement of the first class, and no changes will be permitted without the approval of the program administrator. Class sizes will be limited to 25 with a minimum of 12 preregistered participants needed to conduct the class. Promotional materials and security clearances will be provided by the health promotion department. All class materials will be provided by the vendor.

they are not in a position to have their needs met. They may find themselves over a barrel.

Although formal negotiations do not take place every day in all health promotion settings, there frequently are occasions that call for informal negotiations. Having skills in this area can be a significant asset to managers.

Sole-Source Purchasing

Sole-source purchasing is often used as a means of streamlining purchases that are made frequently. It involves identifying a vendor that can provide the desired products or services. The purchasing procedures are similar, but this is the equivalent of one-stop shopping for certain categories of products. Examples of products and services that may be candidates for sole-source purchases are

- aerobics instructors,
- incentive items,
- physical testing, and
- printing.

Sole-source purchasing offers the benefits of volume discounts. It involves fewer purchase requisitions and purchase orders and simpler invoicing. The savings in time and expense make it worthwhile for managers to consider sole-source purchasing whenever it is available.

The Role of Legal Counsel

The purchase of products and services involves legal documents. The purchase orders used for routine transactions may be developed in house with input from the organization's legal counsel, or they may be purchased from a supplier and contain standard wording that has been approved by the organization's attorney.

Consulting agreements and contracts may give rise to legal complications. Before signing any agreement on behalf of the organization, managers should have the document reviewed by legal counsel.

Some areas of potential risk for health promotion programs are below:

- Liability for injuries that occur during participation in sanctioned program activities
- Copyright infringement by vendors providing materials to the program
- Trademark infringement
- Breach of contract litigation if the terms of a contract are not fulfilled
- Shared liability for injury or damage if products are not used as specified by the manufacturer
- Conflicts of interest

These examples underline the importance of consulting legal counsel before entering into any agreement.

SUMMARY

The purchasing process can range from simple to complex depending on the nature of the purchase and the procedures established by the organization. The process usually involves the use of purchase requisitions, purchase orders, and a vendor selection or bidding process.

Managers can assist in the purchasing process by maintaining current files of vendors and their products and by making recommendations to the purchasing department while respecting their roles and responsibilities. Managers also need to be knowledgeable about writing statements of work, negotiating, sole-source purchasing, using consultants, and legally reviewing contracts and consulting agreements.

The financial aspects of operating a health promotion program are economic conditions, the budgeting process, and the purchasing process. Together they are important for the program's success. However, successful financial operations require effective planning. This is the topic of Part V.

KEY TERMS

purchase order (p. 190)
purchase requisition (p. 189)
request for proposal (pp. 194)

SUGGESTED RESOURCES

Fearson, H. (1988). *Purchasing organizational relationships*. Tempe, AZ: Center for Advanced Purchasing Studies.

Janson, R. (1988). *Purchasing ethical practices*. Tempe, AZ: Ernst & Whinney.

Scheuing, E. (1989). *Purchasing management*. Englewood Cliffs, NJ: Prentice-Hall.

U.S. Small Business Administration. (1984). *Purchasing and controlling costs*. Washington, DC: Author.

Part V

Planning

To manage a health promotion program effectively, managers must develop a comprehensive, detailed plan. The plan should identify objectives and specify the steps needed to achieve them, including the allocation of financial resources; the thrust of marketing efforts; and the hiring, training, and appraisal of employees. In Part V we discuss the basic planning cycle.

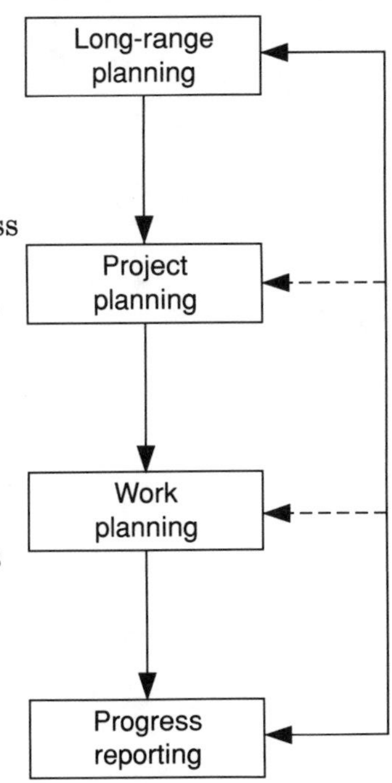

The planning process is a cycle of activities as shown in the figure here. In chapter 21, we discuss the fundamentals of developing a long-range plan. The long-range plan is divided into projects, and a plan is developed for each project; this is the subject of chapter 22. In chapter 23 we explain how to divide the project plan into individual work plans. The cycle is completed with progress reporting, described in chapter 24.

Depending on the results of progress reporting, the long-range plan may need to be modified; this begins the planning cycle again. To attain the goals of the department and organization, it is essential to follow this systematic approach to planning.

Chapter 21

Long-Range Planning

Long-range planning is essential to the success of any organization. It provides a vision of the future and a blueprint for the orderly achievement of well-defined goals. Resources will be used wisely, and managers will have the direction and guidance they need to make critical decisions. For these and other reasons, business operations should be driven by a formal long-range plan.

By its nature, health promotion is a long-range project. Health professionals recognize that behaviors that have developed over years will not change appreciably or permanently over a relatively short period of time. For this reason alone, it is prudent to develop a long-range plan for a health promotion program.

DEVELOPING A LONG-RANGE PLAN

There are a variety of ways in which a long-range health promotion plan can be developed. Ideally, the process would be initiated and directed by a key executive charged with managing the organization's health care claims situation. That person would become the "program champion" for the health promotion effort and would direct that aspect of the organization's long-range plan. There would also be a goal statement and a strategy for the development of a comprehensive health promotion program.

This scenario is ideal but unlikely. It is more probable that a middle manager will be responsible for long-range health promotion planning and may delegate some of the work to subordinates.

The Program Champion

Although a long-range plan can be developed without the support of a program champion, having one will facilitate the process and enhance the result. The ideal champion is a key executive who has a secure position in the organization, the respect of peers, and sufficient influence to help move progressive programming ideas that challenge the norms of the worksite culture through the decision-making and approval process.

The Planning Committee

With or without a champion, the health promotion program and the long-range plan must have a broad base of support. Any employee or department that has a stake in the problem of health care costs, or in any aspect of the plan, should provide input to the plan and support its objectives.

To facilitate this process, a committee should be formed. This can be a planning committee that outlines, structures, and refines the plan or an advisory committee that provides ideas, direction, information, and consensus on the various aspects of the plan. Possible members of the committee are

- the chief executive officer or a delegate,
- the chief financial officer or a delegate,
- the medical director,
- the employee benefits manager,
- the human resources manager,
- the occupational health nurse,
- the health promotion manager,
- the fitness director,
- the employee recreation manager,
- the union leader, and
- any interested employee.

In smaller organizations, one person may be serving in several of these roles and can represent the interests of all of these areas of responsibility. The committee should consist of from five to seven members. If possible, the health promotion manager should chair the committee, because this department has the most at stake.

After a committee has been established, the planning process begins. It consists of data collection, writing a mission statement, writing goal statements, determining evaluation standards, devising strategies, obtaining management approval, and distributing or communicating the plan to everyone who has a role in implementing it.

Data Collection

In gathering data for long-range planning, many of the sources described in chapter 15 on market research can be used. The most useful information will be health claims data, aggregate health risk assessment data, health survey data, and demographic data. The data should be summarized in a format that will be easy for the committee to use. Much of it also may be used as supporting documents for the final plan.

Management Approval

As the long-range plan takes shape, committee members should seek management approvals while the various sections are in draft form. If the project is moving in a direction that will not receive final approval for any reason, it will save time and energy to know that early in the process. Management is more likely to approve the final plan if they have given approvals along the way.

Some political maneuvering is often necessary to get the job done. This is when a program champion becomes invaluable. In any case, the committee should seek written approval of the plan by any manager or executive in the chain of command who has the power to approve significant decisions, especially in the area of resources and policy, after the plan goes into effect.

Once the plan is complete and approved, it should become the foundation for all health promotion activity. Project plans (plans for an individual division, facility, or major activity such as the education program) are all derived from the long-range plan. Subsequently, individual work plans will be based on the project plans. At this point, the commitment developed by having a representative planning committee will pay off, and no one should have to be "sold" on the plan. Their next steps should be clear.

Time Span

The length of time for which a long-range plan is written will vary with the nature of the program. If a program is intended to be a short-term recruiting tool or a perquisite for management, the time span will probably be shorter than that for a plan whose goal is to significantly improve the health of employees and reduce health care costs. In most cases, a long-range plan will cover from three to five years. However, the length may be dictated by factors such as business cycles, financial forecasts, or the organization's overall long-range plan. In this case, management will give the health promotion department direction in the planning process.

The Mission Statement

A long-range plan needs a vision. The vision can be articulated in the plan's mission statement (see box on page 202). Most organizations have a mission statement from which all other activities flow. Similarly, the health promotion department also should have a clear picture of its mission.

The health promotion mission statement must have the support of senior management. This will increase the likelihood of support for decisions that relate to the accomplishment of the mission.

The mission statement should represent the true reason for the existence of the health promotion program. Once it has been completed and approved, planners can turn their attention to the goal statements.

Statements of Goals

The mission statement will contain statements of broad goals without quantification. Examples of such statements are the following:

- Reducing lifestyle-related health care claims
- Creating a healthier work environment
- Improving the quality of life of employees and their families
- Increasing productivity
- Improving morale
- Attracting and retaining high-quality employees at all levels

Such phrases are incorporated into a brief statement that may address several objectives. For example:

The XYZ health promotion department is committed to providing a positive work environment and an improved quality of life by offering employees opportunities to improve their physical and mental health through innovative lifestyle-related health education programs and activities.

Once the general direction of a program has been set by the mission statement, specific operational goals for accomplishing the mission must be established. These goals must

- be specific,
- be measurable,
- be attainable, and
- include clear deadlines for completion.

Mission Statement

The We Care Health Promotion Program has been established to help all employees achieve optimal health. We will accomplish this by providing employees health risk awareness information, health promotion educational programming, and a supportive and healthy environment. The We Care Health Promotion Program will serve all employees and the company by developing employee morale and a healthier work force.

In establishing goals for the health promotion program, planners should have an accurate picture of where the program is now. This will enable planners to ascertain whether a goal is in fact attainable. It also will provide a basis for evaluating progress toward achieving goals.

An example of a goal that meets these criteria is: "By 12/31/99, XX% of the male employee population will have blood pressure of 120/80 or less as measured at their mandatory annual physical." Another example is: "By 9/30/99, YY% of the employee population will use seat belts as measured by a parking lot entrance census."

The fact that no baseline data are available in a given risk area should not prevent planners from setting goals in that area. Planners may set several goals in such a case. The first goal may be to evaluate that area in the first year to establish a baseline, and the second goal may be to develop program ideas in the second year based on the new data. It is always possible to establish new goals during the annual review cycle.

Although evaluation standards may be specified in the goal statement itself, for clarity it may be useful to include separate statements describing the factors to be evaluated. Examples of items to be evaluated are

- total claims for lifestyle-related health problems,
- total participation as a percentage of the eligible population,
- percentage at risk in specific target populations as measured by an aggregate health risk appraisal report or an employee health survey, and
- total number of executives who qualify for bonuses based on program participation by their employees.

Developing Strategies

After goals and evaluation standards have been established, planners must identify strategies for attainment of the goals. Every goal will require at least one strategy. However, one strategy may help achieve several goals.

A strategy must be stated with some specificity. However, the detailed steps for implementing the strategy will be set forth in project plans and individual work plans.

If strategies are to be meaningful, they must be clearly related to goals. For example, a goal of improved management support could have a strategy of linking executives' bonuses to their employees' participation in health promotion programs. The name, title, and department of each person who will be responsible for implementing the strategy should be specified, as should the other departments whose assistance may be required.

Here are some other examples of strategies:

- Forming a wellness committee consisting of representatives from all departments to help develop plans relative to participation goals
- Including nutritional considerations in the contract for food vending services to achieve goals in cholesterol management
- Establishing a "wellness grant" fund to encourage employee-driven health promotion project goals
- Offering a life/automobile insurance incentive for seat belt usage to achieve seat belt compliance goals

Developing strategies is an exciting and challenging planning task where creativity is an asset. Innovative strategies are one key to a highly successful program. The essence of adult health education is attempting to bring new, more effective interventions to a population that in many cases has tried to quit smoking numerous times, lose weight only to gain it back, or make other equally difficult lifestyle changes. Employees who have "tried it all" become the target populations. Only a fresh approach will motivate them to try again.

Traditional brainstorming techniques are extremely useful during strategy sessions. Part of the fun of devising strategies is that the committee does not have to do the hard work involved in implementing them. This becomes the responsibility of the health promotion staff.

Format

The format of a long-range plan can vary. In large organizations, the format may be specified and forms provided on which departments prepare their plans. Lacking a standard, an outline will work very well in most cases. It is important that the plan be easy to follow, clear, and concise. To this end, the various sections should relate logically to each other, especially the goals and strategies. A format outline follows.

I. Mission Statement
II. Current Status
    A. Background information
    B. Employee population risk levels
    C. Employee program participation rates
    D. Other relevant baseline data
III. Program Goals
    A. Goal #1
    B. Goal #2
    C. Goal N . . . .
IV. Strategies
    A. Strategy #1
        1. goals affected by strategy #1
        2. functional responsibility for strategy #1
    B. Strategy #2
        1. goals affected by strategy #2
        2. functional responsibility for strategy #2
    C. Strategy N . . .
V. Evaluation
    A. Methods for measuring behavior change
    B. Methods for measuring participation
    C. Methods for measuring financial impact
    D. Other methods

THE REVIEW AND REVISION PROCESS

The last component of the long-range planning process is the review and revision cycle. In most organizations, senior management dictates the timing of planning. Usually it is tied to an annual budget cycle.

The amount of lead time for planning usually depends on the size of the organization, with larger organizations tending to need more lead time. For example, if the fiscal year begins on January 1, preliminary budget planning may begin in August or September with the first budget to be submitted by early October. This means that managers should begin in July to review their long-range plans and develop forecasts based on progress toward achieving goals.

The revision process involves many of the same steps used in the initial development of the plan. However, it is usually easier and quicker, and the entire planning committee may not need to be involved. It is desirable to obtain input from anyone affected by the revision, but the responsibility for the revision will probably remain with the health promotion manager. The most important factors that will influence the revision process are changes in assumptions, changes in relevant data, the relative success of the various strategies implemented to date, and the progress made toward the existing goals.

A long-range plan may reach the revision stage with a few goals accomplished, a few goals modified in quantity or deadlines, and a few new goals. Strategies that are successful and still needed probably will be retained, but some will have served their purpose and therefore will be eliminated from the revision. If new goals are introduced, new strategies also may be needed. Once the revision is complete and has been approved by management, the process begins again for the upcoming year.

Changing Assumptions

The need to make changes in assumptions can be determined by gathering the latest data, finding out from the human resources department whether any significant changes are forecasted for the population, obtaining sales forecasts and the work schedules they drive, and collecting any other relevant information management will share.

Strategy Review

The extent to which strategies are successful should be clear from activity reports, staff meeting minutes, and memory. Any other documentation kept during the year will also be useful in reviewing strategies. Depending on their success and changes in assumptions, strategies may continue to be used without change, be modified, or may be eliminated entirely. New strategies also may be introduced at this time.

Progress Toward Goals

Progress toward goals will be documented in a manner similar to that used to evaluate strategies. In addition, formal management reports can be used for this purpose. Goals that were achieved may be extended or increased. Goals that were not achieved should be evaluated to determine whether they are realistic and modified if necessary. If there were dramatic shifts in the population's demographics, entirely new goals may be introduced.

Mission Statement Review

Unless there has been a major shift in the organization's management or philosophy or significant changes in internal or external economic conditions, the mission statement should remain relatively intact. Regardless of circumstances, the mission statement should be reexamined regularly to be sure it is still accurate, workable, and supported by all stakeholders. Necessary changes should be made.

BENEFITS OF A LONG-RANGE PLAN

There are a number of significant advantages in having a carefully developed long-range plan for health promotion operations.

It will give direction to the program, enhance return on investment, provide a basis for program evaluation, allow the flexibility to adjust to a changing environment, improve management commitment, and more thoroughly integrate the program into the culture of the organization.

Program Direction

Beginning with the mission statement, the direction of the health promotion program should be dictated by the long-range plan. Because the mission statement is philosophical in nature, the philosophy should permeate all aspects of the program. If management has endorsed a mission statement that specifies a healthier worksite environment, this can have implications for smoking policies, workload patterns, and employee behavior at the Christmas party. It also may influence the relationship between the health promotion and facilities departments or the policy committee. Significant decisions relating to the building of a fitness center as opposed to instituting a paper-based program can evolve from the philosophical position of the mission statement.

The goals identified in the long-range plan dictate the direction of the health promotion program. They are the basis for strategies as well as annual project plans, individual work plans, and ultimately daily to do lists that give direction for actions as basic as making phone calls or writing newsletter articles. If vendors supply all or some program services, clearer statements of work can be written in their contracts to discourage pressure to purchase unneeded services.

Program Effectiveness

Assuming that one purpose of a health promotion program is to control health care costs, a long-range plan will enhance the department's ability to make an impact in this area. For example, if an analysis of claims data indicates that back injuries are the most significant health-related problem, massive smoking cessation efforts will be much less useful than low back risk assessments, appropriate education, and exercise. Such decisions can determine when a program will reach the break-even point.

A long-range plan also will enable health promotion professionals to develop programs to meet the real needs of the organization and the participants. Resources can be allocated more efficiently; budgets, purchase orders, and capital expenditures can be more easily justified; and results can be more efficiently tracked.

Program Evaluation

Because the goal statements in the long-range plan are quantifiable, and the standards for evaluating progress are also specified, evaluation of the overall program is guided by the plan. If there is no evaluation plan, establishing one should be a priority.

Program Modification

With a long-range plan in place, managers should be aware not only of the direction in which they are moving, but also their starting points and the assumptions underlying the goals and strategies. This allows them to be more aware of the impact that changes in their environment, and subsequently changes in assumptions, will have on program activities and decisions.

For example, the acquisition of an organization whose employees are predominantly female could have profound implications for maternity-related health care costs or daycare needs that may relate to health promotion activities. Awareness of these implications enables managers to make program decisions that respond to changes in plan assumptions. Other examples of changes that can affect program decisions are

- divestiture of significant employee populations,
- budget cuts or windfalls,
- reorganization or a new CEO,
- public health issues such as AIDS,
- major changes in an employee benefit plan,
- significant changes in the work cycle such as extended periods of overtime, and
- rumors or the real threat of significant layoffs.

Management Commitment

One of the most significant benefits of a well-developed long-range plan is more sincere and enduring management commitment to the program specifically and to the concept of wellness in general. Senior managers can be expected to support a plan that explains the causes and impact of rising health care costs and sets forth specific strategies for reducing these costs through health promotion activities. Management commitment and contributions to achieving solutions become a valuable resource for the health promotion department.

Incorporating the Program Into the Worksite Culture

One of the hallmarks of a successful health promotion program is that it no longer appears to be a program, but is a way of life in the workplace. The norms of the population change, and the program becomes woven into the fabric of the worksite culture. At this point, it would take extraordinary circumstances to lessen management's commitment to a program that is so important to employees.

SUMMARY

Long-range planning is essential for any health promotion program that has a goal of contributing to an organization's efforts to control health care costs. Whether driven by the organization's overall strategic plan or developed by a health promotion professional or a planning committee, a good long-range plan will give a program direction, improved and quicker return on investment, a basis for evaluation, enhanced management commitment, and the ability to adjust to changing conditions within and outside the workplace.

Built on a comprehensive mission statement, a long-range plan will spell out operational goals, strategies, evaluation standards, roles, and responsibilities. Through the review and revision process, managers can track progress and make adjustments as changing circumstances dictate. In addition, the long-range plan will be useful in subsequent project planning, the next step in the planning process, which we address in chapter 22.

KEY TERMS

program champion (p. 200)
mission statement (p. 202)

SUGGESTED RESOURCES

Drucker, P. (1954). *The practice of management*. New York: Harper & Row.

Ellis, D., & Pekar, P. (1980). *Planning for nonplanners: Planning basics for managers*. New York: AMACOM.

Levin, D. (1981). *The executive's illustrated primer of long-range planning*. Englewood Cliffs, NJ: Prentice-Hall.

Shanklin, W., & Ryans, J. (1985). *Thinking strategically: Planning for your company's future*. New York: Random House.

Chapter 22

Project Planning

After you have a long-range plan in place, project planning is the next step. Though the overall long-range plan gives relatively detailed information in the areas of goals and strategies, it is still too global for operational application. The planning process needs to be taken to yet another level of detail.

WHAT IS PROJECT PLANNING?

Project planning can be structured in a number of ways. In a very large organization, it might be the basic plan for a unit within the organization (such as a department or division). In a geographically dispersed organization, project planning might be done for a facility, plant, city, state, or region. In other situations planning might be for a specific activity, such as health screening for a certain population.

Many traditional project planning techniques can be applied in health promotion settings. Techniques such as Gantt charts, PERT charts, and process flow analysis all are appropriate. However, in many cases they would be overkill, because many projects do not have the complexity for which these sophisticated methods were developed. If you do have a need for such methods, you'll find many computer programs available for applying them easily. (These tools are detailed in the suggested resources listed at the end of the chapter.) In this chapter we'll present a more basic approach using modifications of these techniques.

The essence of project planning is to turn the strategies from your long-range plan into detailed schedules of activities. The project plan describes what will happen, when and where it will happen, who will make it happen, and what resources will be used to make it all happen. Although the project plan flows from the strategy more than from the goals, you need to keep your goals in mind so that your project has the proper focus. For example, if your long-range goal involves reducing cholesterol levels and your strategy involves introducing nutrition education, keeping your goal in mind ensures that any courses offered emphasize cholesterol reduction strategies over other nutritional concerns. At the project level, this plan would include course schedules, specific vendors or instructors, training needs, ordering and distributing materials, some detail on promotional events, and registration information. It would not, however, be as detailed as a work plan or to do list (discussed on pp. 219-224).

INFORMATION COLLECTION

Before you can begin detailing a project plan, you must do a certain amount of information gathering. Information is needed in two main areas: logistical (such as availability of meeting rooms or of shipping and receiving services) and marketing (described on pp. 133-141). As a rule, this information gathering probably will be done in one major effort combining marketing, site surveys, interviews, and other methods. (The areas are treated separately only for the purpose of making their relationship to planning clear.) In the first year of a program you'll emphasize basic logistical information, much of which will remain unchanged in subsequent years.

Logistical Data

Logistical data is necessary for the operational aspects of the project plan. This data can be gathered by means of written surveys, interviews, or on-site inspections. Several kinds of data are needed to plan a project thoroughly.

Personnel

All of the people who can play a key role in a health promotion project and should become part of a variety of support networks include

- secretaries,
- security guards,
- shipping and receiving staff,
- mailroom staff,
- printing/reproduction staff,
- audiovisual staff,
- medical director,
- occupational health nurse,
- safety director,
- first responder (employees trained as EMTs or in CPR),
- cafeteria/food vendors,
- facilities manager and staff,
- transportation manager and drivers,
- line management (especially in manufacturing settings),
- a variety of key managers, and
- employee services/recreation staff.

Facilities

Facilities that are important in planning a project may include

- meeting rooms,
- existing exercise facilities,
- possible exercise rooms,
- cafeteria or other large common space,
- employee entrances and exits,
- bulletin boards,
- petty cash windows,
- copy machines,
- time card punch clocks,

- mailroom,
- laboratories, and
- showers.

Policies

Organizational policies that can help or hinder program activities may include

- smoking policy,
- alcohol policy,
- drug/drug testing policy,
- public relations/communication policy, and
- program-specific policies
  - eligibility
  - program cost sharing or copayment
  - use of company time
  - use of facilities after hours
  - purchasing procedures

Miscellaneous

Other information that will be useful in the planning process includes

- number of shifts in operation,
- number of employees on each shift,
- shift beginning and ending times,
- break and lunch times,
- union-related considerations,
- ethnic considerations,
- considerations of employees who are disabled,
- impending events that could affect program plans such as extended overtime or layoffs,
- local health issues such as excessive stress or AIDS,
- locations of rooms suitable for classroom activities,
- locations of rooms suitable for fitness activities,
- locations of rooms suitable for health screening activities,
- room reservation systems,
- availability and location of audiovisual equipment,
- approved bulletin boards or other posting areas,
- electronic mail capabilities,
- in-house newsletters,
- printing or other reproduction resources, and
- availability of library or learning center services.

Methods of Gathering Information

Three common methods of gathering information for a project plan are written surveys, physical inspections, and interviews.

Written Surveys

A good way to start gathering information is to develop a written survey that covers all of the factors listed above. The survey should be conducted well in advance of the planning cycle with an update just before the writing of the final draft of the project plan. First send the survey to the appropriate people and request their assistance. Then seek an interview with each person to ensure that responses are accurate, clear, and complete.

Physical Inspections

After compiling responses to the survey, managers should make a physical inspection of the facilities to clarify certain characteristics as well as to begin the networking process. While walking through the site, inspect meeting rooms and estimate their capacity for various activities. The person who completed the survey probably indicated room capacity based on standard business meetings. Rooms may be rated differently for health education classes, seminars, exercise classes, or health screening. Not every room will be suitable for health screening activities or physical exercise.

Rooms that might be used for health screening should be carefully mapped out.

Make a rough sketch of each room indicating the dimensions and approximate placement of screening stations and the traffic flow. For ease of moving materials, note the room's proximity to shipping and receiving. Space that might be used for aerobics classes should be evaluated for capacity, ease of moving furniture, and safety features such as carpeting or other padding. Take note of informal employee gathering places where a dedicated posting area could be useful. Throughout this process, use checklists to ensure that information gathered is complete.

During a physical inspection, there are often opportunities to meet people who can be helpful to the project. Making a good first impression and learning something about them will be useful in subsequent meetings where network building begins. Take notes to assist in recalling names, titles, and personal information that will aid in building relationships as support networks are put in place.

Interviews

After gathering information through surveys and physical inspections, the next step is to interview the survey respondents. These meetings should be cordial, brief, and have a definite agenda and goal. Each meeting should end with some kind of commitment from the interviewee: agreeing to endorse the program or activity, co-sign a memo to employees, enroll in the program, be a group leader, become part of a network, or merely agreeing to meet again for further discussions.

NETWORK BUILDING

Networks can be a key element in the success of a health promotion program. Networks can be established within several employee groups, each contributing to the effectiveness of the program.

To keep a network alive and healthy, managers must nurture and reinforce it. Network participants should be rewarded for their contributions with special incentive items not available to other employees or with personal memos of thanks (with copies sent to their supervisors).

Rationale

There are a number of reasons to invest time and energy in building networks in an organization. First, it is a way to augment the resources allocated to the health promotion department. For example, it may be unnecessary to create a newsletter dedicated to health promotion if the editor of the in-house newsletter can be persuaded to publish an insert or column that will accomplish the same purpose. This may be even more effective than a dedicated newsletter if the publication is well accepted by employees.

Second, the most successful programs have a high degree of employee commitment. This is accomplished by giving employees meaningful roles to play and recognizing them for their contributions.

Third, because health promotion is one of the functions that crosses all organizational lines, it is ideally suited as a center for many interconnected networks.

If networks are to play a significant role in the implementation of a health promotion project, it may be necessary to adjust the program schedule to coincide with that of a network member. For example, the announcement of a health education class schedule must be timed to fit into the newsletter's publication schedule. Because deadlines and related activities are significant factors in a project plan, there must be cooperation with network partners.

Employee networks can be of significant help in implementing a health promotion project. Networks can solve logistical problems, improve communication, and extend resources. Having a network of cooperative employees can simplify many tasks. There are a number of departments and classes of employees that can be useful in logistical tasks.

Secretaries

Secretaries often handle functions that are helpful in conducting a health promotion program. Some of these functions are

- reserving rooms,
- reserving and/or setting up audiovisual equipment,
- routing communications within a department,
- posting flyers on bulletin boards,
- providing emergency backup clerical support,
- greeting and escorting visiting speakers or instructors,
- duplicating printed materials, and
- handling registration for activities.

Facilities/Maintenance

The facilities department controls the physical facilities needed to conduct health promotion activities and the labor needed to set up and take down equipment efficiently. (In some organizations, especially those that are unionized, there may be strict rules as to which employees may do certain types of work.) Some of the tasks a facilities department can assist with are these:

- Receiving and delivering equipment and materials to the site of a health promotion activity, especially at remote facilities
- Setting up and/or taking down cafeteria or conference room tables for exercise classes or health screenings
- Troubleshooting unexpected problems at program activities, such as a power failure or equipment malfunction

Mailroom Personnel

Effective communication is essential to the success of a health promotion program. Because communications to employees are either mailed to their homes or delivered at work, the mailroom staff can be a vital asset to the program. Here are some of the ways they can be useful:

- Facilitating priority or last-minute mailings
- Providing up-to-date employee mailing lists (often more quickly than human resources can)
- Providing "special delivery" to the health promotion office when needed

Security

The security department can be a valuable ally of program activities in a highly secured building. With the help of these employees, many activities will run smoothly without delays related to security issues. They often can assist in these ways:

- Providing blanket security passes that allow outside health promotion personnel easy passage with needed materials
- Providing security clearances for visiting speakers, instructors, spouses, or professional colleagues
- Offering the services of sometimes idle receptionist guards for stuffing envelopes, labeling, or folding flyers
- Allowing the desks at facility entrances and exits to be used as locations for literature displays
- Volunteering as CPR or first-aid instructors

Employee Recreation Clubs

Most large organizations have an employee recreation club that organizes activities like sports leagues and company picnics. Many of the club's activities overlap with fitness-related health promotion activities. The club may cooperate in these ways:

- Cosponsoring activities such as a running club
- Offering space in its newsletter or other publication
- Providing health promotion activities in conjunction with company picnics such as healthy food options or a sports tournament

Managers

Management support or the lack thereof is often cited as the reason for success or failure of health promotion programs. To encourage managers' support, health promotion professionals should bring them into the network. As a part of a network, managers can help in the following ways:

- Approve activities that, while part of a legitimate health promotion project, are unusual in the organization's typical workday.
- Influence other managers to support the program.
- Ensure that their employees receive all communications from the health promotion staff.
- Serve in a leadership capacity as needed.

Other

Virtually any employee or employee group can be part of a network that supports the health promotion program. Activity groups such as aerobics classes and running clubs need leaders. These leaders may form a network that can serve in a variety of ways. The program should have an advisory committee, which can become a network that represents a wide variety of employee interests.

Network building is included in the discussion of project planning because the definition of a project plan states that the plan should indicate what will happen, when it will happen, where it will happen, who will make it happen, and what resources will be used to make it all happen. In planning a project, managers should exploit network opportunities, because they may be one of the resources that will be used to make it all happen. The project plan and ultimately the work plan may include activities specifically designed to develop networks.

PROJECT PLAN DEVELOPMENT

After all relevant information has been gathered, the formal project plan should be developed. This document will give specific direction to everyone involved in the implementation of the project. Just as the long-range plan serves as the master document from which the project plan is developed, the project plan will serve as the master document from which individual work plans will be written.

The important factors to consider when developing a project plan are described next.

Format

Several formats can be used for a project plan. The preferred approach is to develop a plan in chronological order. This lends itself to either a simple calendar format or a timeline format, sometimes referred to as a Gantt chart (see Figure 22.1).

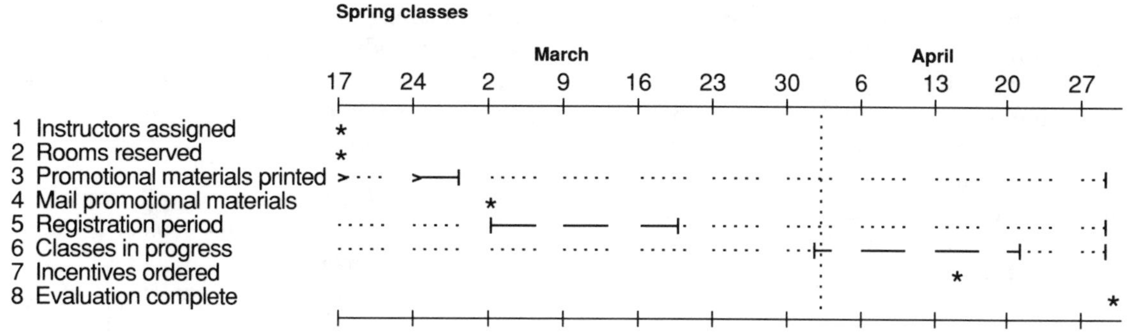

Spring classes

March April

17 24 2 9 16 23 30 6 13 20 27

1 Instructors assigned
2 Rooms reserved
3 Promotional materials printed
4 Mail promotional materials
5 Registration period
6 Classes in progress
7 Incentives ordered
8 Evaluation complete

Figure 22.1 A Gantt chart can visually outline a project and indicate a clear completion date.

Elements of a Project Plan

Project plans contain many interrelated elements: activities, milestones, responsibilities, resources, lag time, and slack time. These terms are common to many project planning systems, including the popular software programs that are available for use with personal computers.

Activities

Activities, or tasks, are the core of project plans. An activity is any event that has defined start and end dates. The time that elapses between these dates is the duration of the activity. Examples of activities are health education courses and health fairs.

A term associated with activities is *predecessor*. This is an activity that must be completed before a subsequent activity can begin. It indicates a dependent relationship that must be kept in mind when laying out a project plan.

Milestones

Milestones are significant events or deadlines. Examples of milestones are major decisions, the scheduled arrival of a piece of equipment, a fitness center grand opening, or the date for a news release or an employee mailing. Milestones serve as checkpoints that may be key indicators for progress or predecessors for activities. For example, class registration should not begin until the class schedule has been announced.

Resources

It is extremely important for health promotion managers to know how their resources are being allocated and whether there is a need for additional resources. The resources of time, materials, money, and labor are the manager's primary tools. For this reason, a project plan usually indicates the resources needed to complete each of the various activities. These resource lists can be used to develop budget line items, personnel headcount forecasts, and purchasing forecasts. A complete listing of the resources needed for a project shows clearly where and when additional resources are needed as well as which resources are being under- or overutilized. Managers then can make the adjustments necessary to complete the project satisfactorily.

Responsibilities

The project plan must indicate who is responsible for completing each activity. These responsibilities are closely related to the resources mentioned above. With responsibilities clearly indicated, employees will have the direction they need to develop work plans.

Implementing a health promotion project may involve working with employees from other departments. In this case the manager must obtain from the managers of those departments the authority to direct the activities of their employees.

Lag Time

Lag time or lead time is the time between the decision to conduct an activity and the time it actually begins. For example, if vendors are used for the printing and mailing of materials to a group of employees, the actual mailing date may be indicated on the project plan as a milestone. Since you are not concerned with the details of the actual execution of the mailing activity, the use of a start time is not appropriate. However, the vendors will undoubtedly need some lead time for their project plan. The vendors' duration becomes the managers' lead or lag time.

Slack Time

Slack time is time within an activity or between activities that is not needed for the activity itself. It amounts to a cushion that can be compressed to adjust the end dates of activities or the completion date of the project itself. Participants can use slack time to complete predecessor activities. Slack time is one of the least expensive contingency planning resources. Often it is built into a project schedule to allow for the inevitable miscues and delays, particularly those that are beyond the manager's control.

Writing the Plan

The actual writing of the project plan begins with laying out the information in an organized fashion. A good way to get a quick, clear picture of the project as a whole is to use a large sheet of paper and a column format. The format shown in Table 22.1 is suggested as a method of organizing the project information:

- Column One - A list of all activities needed to complete the project in chronological order.
- Column Two - The best estimate of the duration of the activity. This can be in hours, days, or weeks depending on the level of detail and magnitude of the plan.
- Column Three - Proposed start dates based on the duration and chronological sequence of the activities.
- Column Four - A list of the end dates.
- Column Five - The amount of slack time if any was included in the duration.
- Column Six - A list of the people responsible for activities.
- Column Seven - The best estimate of the resources needed, excluding time, which is covered by the duration.

This chart is a single source of all the information needed to complete the written project plan. The chart should be carefully examined and revised before the information is transferred to the final project plan.

For purposes of this example, a timeline or Gantt chart will be used. A formal Gantt chart uses standard symbols to differentiate among milestones, activity start and end dates, and other elements of the project. These may be used at the planner's discretion depending on the complexity of the project and the need for clarity. Software packages for planning also use the symbols.

Initial Revisions

Once a project plan has been laid out in a Gantt chart format (or calendar format), planners can begin to analyze relationships

Table 22.1
Project Plan

Activity	Duration	Spring classes Start date	End date	Slack time	Who owns?	Resources
Assign instructors	0	2/17	2/17	7 days	Mgr.	None
Reserve rooms	0	2/17	2/17	7 days	Clerical	None
Print promo materials	5 days	2/24	2/28	0	Clerical	$50
Mail promo materials	0	3/2	3/2	0	Clerical	$100
Registration	14 days	3/2	3/19	11 days	Clerical	0
Classes	3 weeks	4/1	4/21	0	Coord.	
Incentives ordered	0	4/15	4/15	0	Coord.	$75
Evaluation complete	0	4/30	4/30	0	Coord.	0

among activities, responsibilities, resource allocation, and ultimately project completion. At this point the project should not require expensive overtime work or the hiring of supplemental or temporary staff because of inappropriate allocation of labor or unrealistic deadlines. If these conditions are found, revisions should be made and the plan refined until the project is scheduled to be completed in the most efficient manner possible, on time, and on or under budget. This final plan should be distributed to everyone involved and analyzed individually and in project review meetings. Work plans will be developed, and the project can begin.

Ongoing Revisions

Once the project is under way, there are bound to be circumstances that require adjustments to the plan. Managers usually are held responsible only for the completion of a project on time and within the budget. Senior management seldom becomes involved with details of the plan unless important deadlines are in jeopardy or substantial additional resources are needed. Making adjustments to the plan is the health promotion manager's responsibility. Adjustments commonly are made in the allocation of time and labor. Both of these also can affect financial resources.

Time Adjustments

The resource of time is one of the most dynamic variables in a project plan, and it often is one of the few variables a manager can control. For example, assume that the preparation of educational materials for an activity requires two person/days of labor and the duration of the activity is two days. Also assume that there is only one clerical employee assigned to the project, and this person's time is already 50% allocated to other activities during the time this activity is scheduled to take place.

Managers have several options if adjustments are needed. The first option is to adjust the start or end date of the activity to increase the duration. If this activity is a predecessor to another event, moving the end date out also pushes all succeeding dependent activities out. This may or may not affect the overall project completion date. This will be evident on the Gantt chart. Moving the start date up will not affect dependent activities, but it may not be possible if this activity has a predecessor. Again, the Gantt chart will be a useful tool for this analysis.

Although moving either the start or end date for the activity buys additional time for its completion, the change may cause this activity to overlap a different activity in which the clerical employee is scheduled to be involved. If this is the case, it may be necessary to make adjustments in responsibilities.

Labor Adjustments

If in the preceding example it is not possible to make time adjustments, another option is to make adjustments in the area of labor. This will affect the budget for a project, although not necessarily adversely. If the solution is to bring in a temporary clerical employee or request the permanent employee to work overtime, the cost may be absorbed by contingency funds that were built into the plan or offset by reductions in another area of the project. However, another solution could be to use another clerical person that is already on staff and underutilized during the duration of the activity. Permanent staff is usually considered to be a fixed expense. As such, while their labor costs should be allocated to the project, there is no impact on the overall departmental budget. Similarly, it may be possible to assign a salaried (exempt) employee to assist in the activity. Salaried employees generally do not get overtime pay and are often used in this manner. However, as a general management practice, extensive use of exempt employees for overtime work is discouraged. It also may be possible to use bartered or borrowed labor from another department.

Project Review Meetings

As a project progresses, it is important to keep everyone involved in its implementation informed of all changes and the overall status of the project. This can be accomplished with regular project review meetings to discuss problems and develop solutions. Revised copies of the project plan should be distributed and discussed.

Individual progress reports usually are presented in a roundtable format. Managers can meet individually with team members to discuss problems that have no impact on other team members and to discuss individual work plans.

SUMMARY

Project planning, the second of three levels of planning in a health promotion program, is an outgrowth of the long-range plan (the first level of planning). Strategies described in the long-range plan are detailed into activities with start and end dates, resources, responsibilities, and relationships to other activities. These activities are organized chronologically on a chart that allows the manager to easily anticipate problems and make adjustments as needed in the schedule, resources, or responsibilities. During the implementation of a project plan, the project team meets regularly to review and update the plan. The third level of planning is the natural outgrowth of the project plan—individual work plans, which may include work assignments from a variety of projects. Individual work plans are discussed in chapter 23.

KEY TERMS

activities (p. 213)
Gantt chart (p. 212)
lag time (p. 214)
milestones (p. 213)
project planning (p. 207)
resources (p. 213)
slack time (p. 214)

SUGGESTED RESOURCES

Gaynor, A., & Evanson, J. (1992). *Project planning: A guide for practitioners*. Boston: Allyn & Bacon.

Gido, J. (1985). *An introduction to project planning*. New York: Industrial Press.

Goodman, L. (1988). *Project planning and management: An integrated approach for improving productivity*. New York: Van Nostrand Reinhold.

Lewis, J. (1991). *Project planning, scheduling and control: A hands-on guide to bringing projects in on time and on budget*. Chicago: Probus.

Randolph, W., & Posner, B. (1988). *Effective project planning and management: Getting the job done*. Englewood Cliffs, NJ: Prentice-Hall.

Chapter 23

Work Planning

Individual work plans are the third element in the planning process. To develop a work plan, the activities defined in the project plan are broken down into smaller units. Responsibility for these functions is assigned to specific employees. The work plan is a link between the organization's goals and the tasks employees must perform to accomplish those goals.

THE BENEFITS OF AN ANNUAL WORK PLAN

Every employee should develop a work plan that covers the entire year. There are many benefits to an organized work planning system. Once projects are broken down into their final levels of detail, these activities become achievable at the individual level. The efforts of all employees are coordinated under the direction of management to prevent duplication or lapses in responsibilities. Employees become committed to completing their assigned tasks, and this helps ensure the completion of the project. The progress of the project and of individual efforts are documented and communicated to the manager for use in contingency planning and performance appraisals.

Coordination of Efforts

One of a manager's primary responsibilities is to achieve the organization's goals through the efficient use of available resources. Project plans identify the employees who are responsible for specific tasks and establish deadlines. Work plans coordinate these efforts by developing commitment from the people involved.

Commitment to Deadlines and Tasks

Individual work plans are written in the context of the long-range plan and project plans. They are written by the employees who are responsible for completing the project activities. If deadlines specified in the project plan are not achievable, the employee should inform the manager of the problem and suggest solutions. The manager then must make the adjustments needed to accomplish the task. Once agreements are reached on tasks and deadlines, the work plan fosters commitment on the part of employees.

Documenting Progress

Because a work plan is written and reviewed regularly with the manager, progress toward goals is well documented. Managers should have a tickler system so they can talk with subordinates before any deadline passes without satisfactory completion of a task. If problems occur, there is clear documentation of who is responsible for completing the task, and the people involved can discuss solutions. This communication process also is important for contingency planning and performance appraisals.

WORK PLAN FORMAT

Like a long-range plan or project plan, a work plan can be written in several formats. For the sake of consistency, we have used an outline format throughout the section on planning and will use it here. Of course, if senior management specifies a format, that should be followed.

Within any format, several methods can be used to organize material: chronological, functional, or project category. We will describe the chronological and functional methods.

Chronological

The advantage of a chronological format is that it is relatively easy to keep track of deadlines. The disadvantage is that employees may lose sight of the relationships between functions. For example, while employees are developing promotional materials for next quarter's educational activities, they may not be assessing the need to hire and train additional instructors.

One solution to this problem is to include cross-reference notes that indicate any critical relationships that should be examined even though their deadlines are fairly distant. Another approach is to include more detailed levels of activities that move predecessor events into a more prominent position. For example, separate hiring from training and move hiring well ahead of training in the plan.

Functional

When a functional approach is used, the items in any functional category should be arranged chronologically with the deadlines clearly indicated. This allows an employee to glance at any category and focus on the current activities. An example of a functional work plan is shown in Table 23.1.

In a functional organization, the work plan's items are placed in categories that are functionally related. Examples of functional categories are listed below:

- Communication: newsletters, management bulletins, reports, operations reviews, and communication meetings

Table 23.1
Functional Work Plan Excerpts

Projects	Target	Actual
Fitness center committee (on hold)	Ongoing	
Field computer work station setup	6/15	
Complete pilot project	7/15	7/15
Complete final project	8/15	8/15
Education committee (delegated to Mazzeo-Caputo)	Ongoing	
Research project	Ongoing	
Aging committee	Ongoing	
Organ transplant committee	Ongoing	
Host local organ donor coalition meetings	1/29	3/25
Develop company campaign	5/1	5/1
Low back injury task force		
Executive luncheons (delegated to Taggart)		
1. Aging	3/1	3/3
2. Organ transplants	4/12	4/20
3. Cholesterol	7/27	
4. Balance in life	Fall	
5. Personal safety task force	3Q	
6. Remote site delivery pilot testing		
a. HQ preparation	7/1	7/1
b. Supervision delegated to Jones		
c. Development of delivery proposal	9/1	

- Supervision: monthly consolidation of documentation, observation of instructors, one-on-one meetings with employees, and writing or discussion of formal performance appraisals
- Professional development: membership and participation in professional organizations, attendance at conferences, preparation of presentations, and formal education
- Education: scheduling of classes, conducting seminars, arranging for speakers, and promotional activities specific to education but not covered under communication
- Administration: recordkeeping, participant registration, collection of participant fees and copayments, and filing and archiving of medically oriented records
- Finance: budget development, monthly monitoring and forecasting of budgets, and purchasing functions such as letting bids, writing statements of work, and approving invoices

The functional categories used in a work plan will depend on the nature of the organization, project, and position for which the plan is written.

USING A WORK PLAN

Once written, a work plan is an excellent method of organizing an employee's work.

Managers can help make work plans an effective tool by ensuring every employee understands his or her role in the overall plan and by monitoring performance. Employees must discipline themselves to refer regularly to the plan, develop to do lists for daily activity, and make revisions as directed by the manager.

Motivation to Write a Plan

Employees must understand clearly the need for their work plans. First, they should see the relationship of their work plans with the long-range plan and the projects with which they are involved. This will help them understand their contribution as team members to the overall goals of the organization.

Employees also should have access to the manager's work plan. Again, this will help them see the relationship of their efforts to those of their manager. Employees who have subordinates should share their work plans with those people. Further, if employees understand that the manager will be reviewing their performance based on the commitments they made in the plan, they will see the importance of the plan to their career goals.

Individual Review

Employees should be given direction in the use of their work plans, particularly if they are young or new to the organization. The plan should be kept in an easily visible location so it does not become buried by other papers. Ideally, employees should review their work plans weekly, either at the beginning or the end of the week. During the review, the employee should read each section of the plan to bring that week's activities into focus. Notations should be made to document progress and problems. Cross-references to related activities should be reviewed. At the conclusion of this process, which should take only a few minutes, the next week's tasks should be clear.

Translation to Daily Activities

The next step in using a work plan is to translate the activities that should be accomplished during the week into specific tasks. This can be done as a to do list or other appropriate time management tool. For example, a work plan item that specifies communicating the next quarter's course schedule to all fitness instructors could be translated into a list of 10 names and phone numbers to call. It also could be translated into writing a form letter and delegating the mailing of the letter to clerical employees.

At the end of the weekly work cycle, during the next individual review of the work plan, notations are once again made to document progress, and the cycle begins again. The week's accomplishments also should be communicated to the manager in an activity report.

Work Plan Revision

After the manager conducts formal work plan reviews, employees should revise their plans according to the manager's directions. Employees should not revise their work plans without input from the manager, because they may inadvertently delete an important predecessor activity.

MANAGEMENT REVIEW CYCLES

As part of both the project planning process and performance appraisal, managers should review individual work plans at least quarterly and at most monthly. New or young employees initially may need weekly reviews. This is an opportunity for the manager to reinforce good performance, redirect poor performance, gather documentation for performance appraisals, and monitor the project and make appropriate adjustments.

Review Meetings

Regular, formal review meetings should be conducted with each employee. The frequency of meetings depends on the complexity of the project, the maturity of each employee, and the need for direction. Monthly or quarterly schedules are common. It is advisable to schedule review meetings close to one another so the manager can easily make adjustments to aspects of the project plan that involve the efforts of several employees. The manager should have a copy of each employee's work plan in hand, and the project plan should be easily accessible.

The purpose of the meeting is to review all tasks that should have been accomplished since the last review, make revisions as needed, reinforce good performance, and give further direction to improve poor performance. Both the manager and the employee should take notes and should be in full agreement when the meeting ends.

Relationship to Project Plans

As specific activities are discussed, the manager should note individual accomplishments in relation to the overall project plan. Predecessor events are particularly important because the success of the project and the efforts of other employees may depend on these activities. During this discussion, the manager has an opportunity to make adjustments to the project plan and subsequently to the individual work plan. In some cases these adjustments may have to be delayed until all employees have had their review meetings. After all data are gathered, the manager may want to schedule a project review meeting to announce or discuss changes in the project plan.

Project Plan Adjustments

The information obtained during work plan reviews is essential for monitoring a project. Flaws in the plan will become evident. Deadlines that have passed will indicate the need for changes. Deadlines that were achieved early can create slack time or affect the scheduling of activities. The complex relationships between employees' work plans and the project plan can be coordinated in project review meetings.

Adjustments typically are made in deadlines, lead or slack time, and resource allocation, as described on pages 213-216. Constructive solutions may involve adding resources, shifting responsibilities, or adjusting deadlines. Some of these decisions can be made and communicated during the work plan review meeting. All changes should be communicated to the entire staff during the project review meeting. It is essential for all members of a team to understand their roles and the relationship of their efforts to those of other employees.

Performance Management

The work plan review meeting can be a useful tool in performance management. As noted earlier, this is an opportunity to provide feedback to the employee and to reinforce good performance.

One of the most difficult tasks for many managers is to give negative feedback about poor performance. The work plan review meeting is an ideal time for such feedback. The manager first should ask the employee to explain the reasons for a missed deadline or unacceptable performance, and then should turn the discussion to suggestions for improvement. The manager should remind the employee of his or her commitment to meet deadlines and should specify the consequences of failing to complete tasks satisfactorily and on time. Barring unusual circumstances, employees should be held accountable for their performance. Performance management is discussed in more detail in chapters 5 and 6.

Although the work plan with the manager's notations will serve as documentation for performance management purposes, additional documentation may be advisable. Details such as the circumstances surrounding a performance issue as well as employees' reaction to praise or especially to criticism should be noted. Both formal and informal documentation are critical in substantiating performance appraisal discussions, performance improvement plans, and disciplinary actions. It is also essential for supporting salary decisions, promotions, and other rewards.

SUMMARY

The use of individual work plans allows managers to move from the broad perspective of a long-range plan, through the more detailed project plan, down to the day-to-day activities of staff members. It gives the manager more control over the completion of important projects and gives individual employees clearer direction.

Regular review cycles improve communication both upward and downward among staff members. They give managers the information they need to make adjustments in the project plan as well as individual work plans. Reviews facilitate a more coordinated effort by the team.

Performance management is largely centered around the individual work plan. Based on reviews of employees' efforts, managers can reinforce good performance, identify the causes of problems and missed deadlines, take corrective steps, and make adjustments to work plans. Managers should document these discussions to support disciplinary actions, salary and promotion decisions, and other performance-related issues.

Once project and work plans have been developed and are being implemented, managers must report their results. This final step in the planning process is called progress reporting and is addressed in chapter 24.

KEY TERMS

work plan (p. 219)

SUGGESTED RESOURCES

Blanchard, K., & Johnson, S. (1982). *The one minute manager*. New York: Morrow.

Morrisey, G. (1988). *The executive guide to operational planning*. San Francisco: Jossey-Bass.

Vella, C. (1988). *Improved business planning using competitive intelligence*. New York: Quorum Books.

Chapter 24

Progress Reporting

An integral part of the planning process takes place after planning has been completed and plans are being implemented. Progress reporting serves all levels in an organization from senior management to clerical and support employees. It can be accomplished by a number of methods, formats, and schedules. Employees at all levels will be more likely to accept the need for progress reporting if they understand its value to them as individuals and how it strengthens the overall planning process.

PURPOSE

Although senior managers, department managers, and nonmanagement employees need progress reports for some of the same reasons, each of these groups also has some unique needs. Employees should consider these differences when writing reports for their managers, peers, or subordinates.

Senior Management

The higher one goes in an organization, and the larger the organization, the less important details become and the more important relationships to the "big picture" become. In many organizations, the health promotion program is only one of many departments about which senior management is concerned. Their perspective and information needs are broad, and they tend to focus on quantifiable data known as key indicators. Examples of key indicators in the health promotion area are

- employee headcount,
- total participation numbers or percentages by location or program area,
- year-to-date expenses and revenues by major category, and
- percentage of employees at risk in a target area.

Typically, the information upon which key indicators are built is fed into the reporting system from lower levels and is synthesized with other information to yield the final data. Alternatively, in larger organizations many key indicators may be tracked and reported through human resources database systems that allow for easy retrieval of information. Areas of concern are budgets, goals, and relationships between departments or programs.

Budget Implications

Senior management has overall responsibility for the organization's bottom line. This can mean meeting stockholders' profit expectations or ensuring the solvency of a nonprofit entity. For this reason, progress reports to senior management routinely contain information about year-to-date revenues and expenses as well as projections for the next reporting period. Budget information can be communicated in charts and graphs with tables available for backup.

As a general rule, managers at any level do not like surprises. If a department or individual is not likely to stay on target for preapproved financial commitments, this information and the circumstances surrounding it should be transmitted upward as soon as the picture is clear. Even if included in a report, requests for additional funding or headcounts usually must be handled individually and justified at a later time. Managers at higher levels usually are more receptive to problems in this area if the manager reporting them also recommends solutions.

Progress Toward Goals

Progress toward goals is another area of interest to senior management. Day-to-day work plan goals are of little interest. Progress toward broad goals related to the long-range plan is appropriate to address in reports to senior management. Of particular concern are goals that are not likely to be attained. As with budget problems, the reporting manager should provide a brief explanation as well as proposed corrective actions.

It may be useful to have some background information about the individuals who will be receiving the progress report. The reporting manager can highlight progress toward goals relating to areas of special interest to a senior manager. For example, if a senior manager went out on a limb to support the capital expenditure for a new fitness center, the level of participation in the center may be important politically to that executive. An ex-smoker who is now a zealot may want to know how smoking cessation programs are doing.

Managers at this level also are likely to scrutinize the relationships between progress toward goals and budget results. Knowing this, the reporting manager should be prepared to explain any discrepancies.

Relationship With Other Departments

The handling of progress reports by other departments will differ depending on the size of an organization. In smaller organizations, the manager of the health promotion department may have related responsibilities and may actually manage the functions of several departments. For example, an occupational health nurse who has responsibility for workers' compensation and health promotion may report the incidence of back injuries and the status of the back education program. A personnel administrator may report on absenteeism and the impact of health promotion programs.

In larger organizations these departments are separate; department reports should be prepared separately but should make reference to overlapping areas and acknowledge the contributions of others. The reporting manager should point out opportunities for synergism and cost savings. It is useful to keep communications open with colleagues in other departments. Occasionally it may be appropriate to prepare a report jointly.

Department Managers

Although department managers such as the health promotion manager must keep the larger picture in view, they are more involved with the day-to-day details of operations and consequently may require more detailed reports. Reports from subordinates should contain information that department managers will synthesize and then transmit to senior management. The health promotion manager should ascertain how much information and detail is needed by senior management and communicate these needs to subordinates. Information needs may include budget details, functional problems and issues, staff performance, and required documentation.

Budget Details

At the department level, regular information on the status of the budget in detail is vital. Less detail is required if the budget is on track. However, if there is a budget windfall or shortfall in the forecast, it is the manager's responsibility to deal with it. To make effective decisions, the manager should know the details about the deviation from budget. These details may include

- source of the problem,
- size of the problem,
- impact on project goals, and
- recommendations from staff.

The report should contain enough detail to clearly present the issues, implications, and recommendations. Budget details usually can be communicated more effectively in tables than in charts and graphs because specific numbers can be identified more quickly.

Problems and Issues

Although professional-level employees are expected to handle routine problems within their scope of authority, the manager may want to know that they occurred and were appropriately handled. This information allows the manager to keep in touch with the program without getting bogged down in details. For example, knowing that there have been calibration problems with body composition equipment may be important for future equipment purchasing decisions. Employees can solve the problem to keep the flow of work moving and then report briefly to the manager.

Serious problems that are beyond subordinates' scope of authority should be reported so the manager can give advice or intervene. For example, if there has been a series of injuries in an activity, issues of vendor or instructor quality, safety regulations, insurance, and liability may be involved. This certainly would require the manager's intervention.

Performance Management

As detailed in chapter 8, department managers must monitor employee performance. They gather information from direct observation, the observations of others, key indicators, and employee progress reports.

Everyone involved in a health promotion program benefits when the manager has accurate information about employee performance. It is in the organization's interest to have a well-run program and to have documentation that provides legal grounds for decisions about transfers, promotions, and terminations.

Employee progress reports give health promotion managers information they need to manage performance, from day-to-day coaching to employee ranking and salary decisions. At this level, it is imperative to have documentation to support decisions and recommendations in case they are later called into question.

Nonmanagement Employees

At the bottom of any chain of command, nonmanagement employees often consider progress reporting to be an unnecessary and time-consuming task. Employees are more willing to prepare progress reports when they understand how they themselves can benefit from the reporting process. One purpose of a progress report to a superior is to prevent major problems that would require support from that superior. A progress report also documents performance to ensure an accurate performance appraisal and the rewards the employee has earned through good performance.

Preventing Major Problems

As mentioned earlier, managers generally do not like surprises. To manage complex situations, they must have advance knowledge of significant potential problems. Problems that have been satisfactorily handled can simply be documented as appropriate. However, there are several problems that superiors should be informed of immediately. These include

- problems that significantly affect budget forecasts with no obvious solution,
- problems that affect other departments or functions that are within the superior's scope of responsibility,
- problems that have potential for legal liability,
- problems that appear to be escalating after initial attempts to solve them, and
- problems that could be politically damaging to one's superiors or the organization.

Managers should emphasize to employees that reporting such problems is not a sign of weakness. It demonstrates employees' ability to see beyond their own domain and shows an understanding of the larger needs of an organization. It shows trust in the relationships among various levels of an organization and respect for one's superiors. By giving management advance notice of significant problems, employees also protect themselves from being given the entire blame for a problem that was beyond their control.

Documenting Performance

Progress reporting is an ideal forum for "tooting your own horn." It is an opportunity for employees to communicate upward their significant accomplishments as documentation of superior performance. By taking a proactive approach to reporting, employees can minimize the chances of inaccurate performance appraisals, being passed over for advancement or desirable assignments, or being blamed for poorly performed activities.

In large organizations, it is likely that managers with many subordinates will have proportionately less information about the performance of each employee. In widely dispersed organizations, there may be only infrequent face-to-face contact between managers and employees. Regular, detailed progress reports can go a long way toward improving communication.

RELATIONSHIP TO PLANNING

Progress reporting completes the circle of communication in the overall planning process. The process began with information gathering, which led to goal setting. The goals drove the activities of project and individual work plans. The reporting function provides updated information that allows for adjustments in goals and individual activities. The most important relationships of progress reporting are with the long-range plan and with individual work plans.

Relationship to the Long-Range Plan

The long-range planning process was described in chapter 21 as an iterative process. Progress reporting to senior management provides direct feedback on the status of the goals that are of the greatest interest to the organization. Although these status reports may not affect day-to-day activities on a short-term basis, they will influence major decisions as the year proceeds. For example, if a major program effort such as a fitness center expansion was part of the long-range strategy, significantly lower than expected use of the existing facility may cause management to move the project date ahead or even postpone it indefinitely. This in turn could lead to some redirection of promotional activities or incentive plans. The timing of these events may span months rather than a few days or weeks as in work planning. Feedback on aspects of the long-range plan usually takes the form of an operations review with emphasis on the big picture through the use of key indicators.

Relationship to Work Plans

The progress reporting process has a major impact on individual work plans. Because an individual's to do list emanates from the work plan, the work flow will fluctuate over fairly short time spans based on employees' progress or lack thereof as reported to their supervisor. Work plans generally deal with short-term goals and small-scale activities. It is relatively easy to change direction in these areas when the need arises.

The progress report may evoke a quick verbal or written response from a supervisor that will give an employee some specific direction that can be implemented immediately. For example, if the progress report indicates that enrollment in a stress management class is three short of the minimum five days prior to the deadline, the supervisor may instruct the employee to publicize an extension of the deadline, offer a recruitment incentive to those already registered, or hold a lunchtime promotional event. Any of these strategies could be implemented with only a day or two of lead time. In some settings, the employee would already have identified these alternatives. The progress report would simply list them and request a decision from the supervisor.

PROGRESS REPORTING FORMAT AND FREQUENCY

In most organizations and departments there are specified formats or guidelines for progress reports. Any format is acceptable as long as it allows for clear communication of information. Common formats are bulleted lists, narratives, executive summaries, and tables, charts, and graphs.

The frequency with which progress reports are prepared depends on the needs of the organization and of the people to whom reports are submitted. Reports are seldom more frequent than weekly and seldom less frequent than monthly. Quarterly reports may be requested, but usually they would be supplemental. A quarterly report may emphasize key indicators rather than anecdotes and day-to-day events.

The advantage of weekly reports is that the recipient has current information about significant activities. Documentation for the entire year will be more thorough under a weekly system. The obvious disadvantage is the increased paperwork involved in preparing reports. The converse is true of monthly reports, where the danger is in losing valuable documentation. A way to keep track of events and conversations that occur between written reports is to maintain an informal file of notes about incidents and discussions. This information may or may not be included in the next written report.

Managers should consider their own needs as well as the time of their employees when setting a reporting schedule for the department. These needs may change over time. During the startup phase of a new program, there may be so many opportunities for things to go wrong that weekly reports may be needed. New employees may be instructed to submit weekly reports to allow the manager to monitor performance closely during the training period. The reporting needs of the manager's superior also may affect the schedule.

SUMMARY

Progress reporting is a structured communication process within organizations that completes the loop between planning and the daily flow of work. Different levels of management have different needs for information. The most common needs are

- documentation of problems,
- documentation of employee performance,
- updates on the status of the budget,
- information about progress toward goals,
- status of key indicators, and
- anecdotal information.

Progress reports can be written in such formats as

- bulleted lists,
- narratives,
- executive summaries, and
- charts and graphs.

Progress reports may be transmitted with varying frequency in writing or orally at operations reviews. The entire process can be enhanced by sharing reports upward and downward as appropriate and providing prompt written feedback.

KEY TERMS

key indicators (p. 225)

progress reporting (p. 229)

SUGGESTED RESOURCES

Brown, L. (1985). *Effective business report writing*. Englewood Cliffs, NJ: Prentice-Hall.

Hochheiser, R. (1985). *Don't state it—communicate it!: How to put clout in your letters, memos, reports, and proposals*. Woodbury, NY: Barron's.

Lambert, S. (1986). *Presentation graphics on the IBM PC: How to use Microsoft to create dazzling graphics for professional and corporate applications*. Bellevue, WA: Microsoft Press.

Sussams, J. (1991). *How to write effective reports*. Brookfield, VT: Gower.

GLOSSARY

accrual—Money set aside for the future payment of an obligation that is known to exist and can be documented.

acquisition—The purchase of one business by another; can be done in either a friendly or hostile manner.

activity—The smallest individual unit in a project plan. Activities have a specific duration and require resources to be completed.

affiliation—The need for companionship and affection from others; the third level of Maslow's Hierarchy of Human Needs.

application—Computer software designed to run on specific computer hardware. See also *software*.

authority—Influence inside an organizational unit. The probability that the will of an individual will be carried out despite resistance.

autocratic—A leadership style characterized by controlling behavior.

batch processing—Running a group of forms, records, or documents through a computer program or printer all at one time.

benchmarking—A method of comparing one organization or one component of an organization with another organization or component that is considered a leader in the field.

bottom-up budgeting—A budget development process that begins with requests from the lowest levels of an organization.

budget—A formal, structured financial document that specifies expenditures and sources of revenue for a specific period of time.

burden rate—A rate expressed as a percentage of labor dollars that covers all services and activities that support the actual production of a product. Burden usually includes staff, accounting, and other indirect costs.

bureaucratic—A leadership style that relies on structure, policies, and procedures.

cell—An individual section on a computer spreadsheet designated by column and row numbers and/or letters in which data are entered and displayed, analogous to a square on a piece of graph paper.

clique—An informal group of employees with similar interests.

competition—Anything that a consumer can choose to do instead of participating in the health promotion program.

consumer—The individual end user for which program activities are planned and delivered. See also *customer* and *end user*.

customer—The recipient of a product or service. See also *consumer* and *end user*.

delegation of authority—The process of authorizing employees at lower levels in the organization to assume specific responsibilities.

demand—Market forces that consume products or services, thereby reducing supply.

democratic—A participative leadership style that allows input from the group.

demographics—Characteristics of any group of people that can be used to categorize them statistically. Common demographic characteristics are age, sex, and educational level.

directive interview—A highly structured interview with specific questions in a specific order.

dissatisfier—A factor that causes employees to have negative attitudes about their jobs, which if absent do not necessarily result in positive attitudes.

divestiture—The process of separating a business entity from a parent organization. This is usually accomplished by sale or management /employee buyout.

downsizing—General work force reduction as a result of lower demand for products or services or the need to reduce operating costs.

duration—The length of time needed to complete an activity in a project plan expressed in working time.

end user—The customer who is at the end of the distribution chain for a product or service. See also *consumer* and *customer*.

expectancy theory—The theory that motivation depends on the belief that a specific action will be followed by a specific outcome.

field—A cateogry of information in a database. A field can take the form of text, numeric data, logical statements, dates, or other data.

fixed cost—A cost that remains constant regardless of production or consumption. Examples are occupancy and capital equipment.

font—A style of type.

forecasting—A repetitive process in the budget cycle where year-to-date expenses and revenue are tabulated and the remainder of the fiscal year's expenses and revenue is estimated. Usually done monthly in business environments.

functional organization chart—An organizational chart structured around similar activities.

Gantt chart—A visual portrayal of a project that includes the start and end dates of all activities as well as milestones. It is useful to show the flow of a project as well as the dependencies between activities.

goal—A statement of expected achievements as the result of execution of a strategy. Reasonable goals are measurable, attainable, and objectively stated with clear deadlines for completion.

grapevine—An informal structure in organizations by which information is passed from one individual to another. The most common method of communicating rumors.

hardware—Computer system components, including the central processing unit, monitor, printer, disk drive, and other electronic and mechanical components.

indemnity plan—An insurance plan where the premiums of an entire group are collected in anticipation of the need to pay the claims of a portion of the group. The risk is spread over the entire covered population, and premiums are readjusted based on actual experience.

inflation—A situation in which prices rise because the volume of money and credit is increasing faster than amount of available goods and services.

integration need—The need for consistency; the fifth level in Maslow's Hierarchy of Needs.

interface—The method by which computer hardware is connected and is able to communicate. This term has been adapted to a human environment to denote one person interacting with another.

intraorganizational segmentation—The process of analyzing an organization and dividing the employees into logical groups of like characteristics relevant to the delivery of health promotion programs.

job description—A written statement of the duties associated with a specific job.

job type—A general category of employees with similar job functions. Examples are management, clerical support, and sales.

job specifications—The skills and knowledge needed to perform a specific job.

key indicator—A specific measurement selected by management from all available data to be reported on a regular basis. Key indicators summarize the factors deemed most important to describing the progress of a project.

lag time—The amount of time one must wait before an activity can begin; often called lead time.

laissez-faire—A leadership style that allows the members of the group to act.

lead—In the sales process, the name of a person or group that may purchase products or services.

leverage—The process of accessing resources from both within and outside the organization.

line—Functions directly involved with the production or delivery of a project or service.

management information system (MIS)—Computer software designed to store, process, and report data useful in management decision making.

market segment—A specific group of potential customers that have similar characteristics that allow a common marketing and sales approach to be applied.

matrix organizational chart—An organizational chart organized around many small ongoing projects.

merger—A friendly transaction in which two or more separate business entities combine to form a single entity.

merit rating—A formal process in which supervisors rate employee performance.

milestone—Significant point in the project development process; an informational marker that does not affect scheduling.

mission statement—A statement that summarizes the primary purpose of an organization's existence and philosophy; the overall goal of a long-range plan.

mixed costs—Costs that can be fixed or variable.

MS-DOS—The disk operating system designed by Microsoft Corporation and used to run IBM and IBM-compatible computer systems.

negotiation—A process in which parties reach agreement through bargaining.

nondirective interview—A relatively unstructured interview with open-ended questions.

nonexempt employee—An employee who is subject to fair labor practice laws, is eligible for overtime pay, and usually is paid hourly.

performance appraisal—A formal process of evaluating employee performance.

performance standards—The quantitative and qualitative criteria in a job description.

periodic evaluation and review technique (PERT) chart—A flow chart that shows the order and interrelationships between activities in a project plan; also called a network chart.

policy—A guide to decision making for situations that are expected to recur.

position—How a product or service is intended to be perceived by consumers; one of the Ps of marketing.

power—Influence outside an organizational unit.

price—Value given in exchange for a product or service; one of the Ps of marketing.

procedure—A specific course of action that helps employees implement policies.

product—Tangible good that is manufactured for sale and use by customers; one of the Ps of marketing.

program champion—An executive, usually at a high level, who takes an intense personal interest in the health promotion program and advocates for it to his/her peers. Also called a corporate champion.

progress report—A written report to management that summarizes an employee's activities and details progress toward goals during a period; also called an activity report.

project organizational chart—A chart organized around a few relatively short temporary projects.

project plan—A formal document/procedure used to organize the activities and resources needed to complete a project.

promotional campaign—A series of related events that are designed to induce employees to participate in an activity or event.

prospect—A lead that has been qualified as a viable potential customer.

public relations—A function that is responsible for communication to the public and external organizations.

qualifying information—Criteria applied to leads to determine if they are viable prospects.

recession—A period of reduced economic activity and low consumer confidence.

record—A unit in a database to which a series of data fields will be associated.

reorganization—The restructuring of internal operations and management responsibilities.

resources—The money, time, labor, equipment, and materials needed to complete an activity.

request for proposal (RFP)—A formal request for vendors to submit proposals for providing products or services; defines the customer's needs and specifies a format for the response.

satisfier—A factor that causes employees to have positive attitudes about their jobs, which if absent does not necessarily result in negative attitudes.

segmentation—The process of dividing a market into smaller units with defined common characteristics.

self-actualization—The reaching of one's potential; the highest level on Maslow's Hierarchy of Needs.

separation—The ending of employment by retirement, resignation, termination, or layoff.

service—An intangible designed to provide a benefit to the end user.

slack time—The "cushion" of time within an activity that allows the activity's start date to be adjusted without affecting the project's completion date.

software—Computer programs stored on floppy or hard disks, used to accomplish a variety of tasks. See also *application.*

staff—Positions that support line functions.

strategy—A plan that guides decisions that lead to the attainment of goals.

supply—The amount of products and/or services available at any given time; a market force that interacts with demand.

top-down budgeting—A process in which higher levels of management determine the amounts of money to be allocated to various departments.

total quality management—An integrated management system with the goal of customer satisfaction.

variable cost—A cost that changes with the amount of volume or activity.

work plan—A document in which an employee outlines proposed activities for a defined period of time.

REFERENCES

American Heart Association. (1992). *1993 heart and stroke facts statistics*. Dallas: Author.

Bezold, C., Carlson, R., & Peck, J. (1986). *The future of work and health*. Dover, MA: Auburn House.

Golaszewski, T. (1992). [A job function analysis of worksite health promotion managers]. Unpublished raw data.

Health Care Financing Administration, Office of the Actuary. (1988). Expenditures and percent of gross national product for national health expenditures, by private and public funds, hospital care, and physician services; calendar years 1960-87. *Health Care Financing Review*, **10**, 2.

Herzberg, F., Mausner, B., & Snyderman, B. (1959). *The motivation to work*. New York: Wiley.

Hunt, V. (1993). *Managing for quality: Integrating quality and business strategy*. Homewood, IL: Business One Irwin.

Maslow, A. (1970). *Motivation and personality*. New York: Harper.

Miller, C., & Tricker, R. (1991). Past and future priorities in health promotion in the United States: A survey of experts. *American Journal of Health Promotion*, **5**, 360-367.

Public Health Service. (1990). *Healthy People 2000: National Health Promotion and Disease Prevention Objectives* (PHS Publication No. 91-50212). Washington, DC: U.S. Government Printing Office.

Robert, H. (1989). *Robert's rules of order*. Nashville: T. Nelson.

Russell, R. (1975). *Health Education*. Washington, DC: National Education Association.

Spencer, G. (1986). *Projections of the Hispanic population: 1983-2080*. Current population reports, population estimates and projections. (Series P-25, No. 995.) Washington, DC: U.S. Department of Commerce, Bureau of Census.

Spencer, G. (1989). *Projections of the population of the United States, by age, sex, and race: 1988 to 2080*. Current population reports, population estimates and projections. (Series P-25, No. 1018.) Washington, DC: U.S. Department of Commerce, Bureau of Census.

Taylor, R. (1991). Trends to watch in the 1990's. *Leadership*, 15-18.

Travis, J., & Ryan, R. (1988). *Wellness workbook*. Berkeley, CA: Ten Speed Press.

Vroom, V. (1964). *Work and motivation*. New York: Wiley.

Index

E

F

T

About the Authors

Dr. Bradley Wilson

Timothy Glaros

Dr. Bradley Wilson is an associate professor of health promotion at the University of Cincinnati and graduate coordinator for the health promotion and education program. He holds a PhD in exercise science and an MBA in management from Michigan State University.

In writing this book, Dr. Wilson has applied a wide range of personal experience in the corporate world, in cardiac rehab and health promotion, and in teaching. Plus he has written over 25 professional publications on health promotion topics and has presented papers throughout the United States, Canada, Europe, and Asia.

Dr. Wilson is a fellow and past board member of the Association for Worksite Health Promotion (AWHP) and a member of the Association for the Advancement of Health Education and the National Association for Sport and Physical Education.

Timothy Glaros, owner of Creative Business Consulting, is marketing manager for EAR, an employee assistance program marketed by Ceridian Corporation (formerly Control Data Corporation), in Minneapolis, Minnesota. He is responsible for marketing, product development, and sales support.

In previous positions Glaros was a manager for Control Data's Stay Well program in projects ranging from 800 to more than 35,000 eligible employees, achieving a participation rate greater than 60%. He has managed programs with operating budgets of more than $4 million and introduced strategies that reduced these budgets by 50% in two consecutive years while maintaining the integrity of the programs.

Glaros received his master's degree in education from the University of Minnesota in 1975. Before entering the business world, he taught health and physical education in the Mounds View (Minnesota) Public Schools.

Glaros has been an active member of AWHP since 1983, serving as a regional president and being selected as a fellow in 1991. He is a frequent speaker at conferences throughout the country.